Basic College Mathematics
for the
21st Century

Made Fun and Easy to Learn

Edward L. Green

THOMSON
™
CUSTOM PUBLISHING

Editor: Michael O'Brien
Publishing Services Supervisor: Christina Smith
Manufacturing Supervisor: Garris Blankenship
Marketing Manager: Sara L. Hinckley

Printed in the United States of America

Thomson Learning Custom Publishing
5191 Natorp Blvd.
Mason, Ohio 45040
USA

For information about our products, contact us:
1-800-355-9983
http://www.custom.thomsonlearning.com

International Headquarters
Thomson Learning
International Division
290 Harbor Drive, 2nd Floor
Stamford, CT 06902-7477
USA

UK/Europe/Middle East/South Africa
Thomson Learning
Berkshire House
168-173 High Holborn
London WCIV 7AA

Asia
Thomson Learning
60 Albert Street, #15-01
Albert Complex
Singapore 189969

Canada
Nelson Thomson Learning
1120 Birchmount Road
Toronto, Ontario MIK 5G4
Canada
United Kingdom

Visit us at www.e-riginality.com and learn more about this book and other titles published by Thomson Learning Custom Publishing

ISBN 0-759-31294-X

The Adaptable Courseware Program consists of products and additions to existing Custom Publishing products that are produced from camera-ready copy. Peer review, class testing, and accuracy are primarily the responsibility of the author(s).

PREFACE

The book was written to help college students pass College Assessment Examinations, as well as develop the necessary skills needed to pass college mathematics courses. The topics covered include: arithmetic, algebra and basic geometry.

Each section in this book relates to a different concept. Once a student mastered a particular section, he should go on to the next section. The student must continue to review the sections mastered so he would not forget the material learned.

This book contains two arithmetic examinations as well as two complete Mathematics Skills Assessment Tests. At the end of each examination appears a list of the answers. After studying the appropriate material, the student should self-evaluate his progress by administering to himself these examinations. Homework in the back of the book will reinforce materials taught. Each homework section correlates with a section in the book.

Most colleges have mathematics laboratories. If a student is still having difficulty, be should go with the book to the mathematics laboratory at the college he attends in order to obtain extra help.

ACKNOWLEDGMENTS

This book is dedicated to the encouragement and support my family has given me through the years. During my childhood my mother, Fae Sarah Wigder Green and my brother, Dr. William Green, encouraged me to pursue my education. My father, Jack Green, worked long hours to support his family. Furthermore, my aunts, Rose Saldinger, Molly Wigder, and Ida Wigder assisted me to obtain my goals. My brother-in-law, Jose Chavin, gave me technical assistance.

I am very fortunate in that my beautiful wife, Zelma, always encouraged me to pursue my career. While I was working late at night, she graduated from Brooklyn College with her Masters Degree in Spanish Literature while taking excellent care of our three wonderful sons Elliot, Philip Steven and Seth Andrew.

Contents

Part I: Arithmetic

1. Basic Skills: Addition, Subtraction, Multiplication, Division

A. Addition

Sum: The result of addition.
Addend: A number being added.

Example 1: $349 + 568$

Solution: Arrange in columns and carry the appropriate numbers.

100's	10's	1's
3	4	9
5	6	8
9	1	7

Answer: 917

B. Subtraction

Difference: The result of subtraction.
Subtrahend: The number to be taken from another.
Minuend: The number from which another is to be subtracted.

Example 1: $415 - 289$

Solution:

STEP 1:   Borrow from the 10's column to increase the 5 to 15.

$$
\begin{array}{r}
{\scriptstyle 0\ \ 15} \\
4\ \cancel{1}\ \cancel{5} \\
-\,2\ 8\ 9 \\
\hline
6
\end{array}
$$

STEP 2:   Borrow from the 100's column 10 tens.

$$
\begin{array}{r}
{\scriptstyle 3\ \ 10\ \ 15} \\
\cancel{4}\ \cancel{1}\ 5 \\
-\,2\ 8\ 9 \\
\hline
1\ 2\ 6
\end{array}
$$

Answer: 126

C. Multiplication

Product:	The result of multiplication.
Multiplicand:	The number to be multiplied by another.
Multiplier:	The number by which another is to be multiplied.
Factor:	Two or more numbers being multiplied.

Example 1: 327×289

Solution: Write in columns.

```
      327
    ×289
     2943
    26160
    65400
    94503
```

Answer: 94,503

D. Division:

Quotient:	The result of division.
Divisor:	The number doing the dividing.
Dividend:	The number being divided.
Remainder:	The number left over after completing a division problem.

Example 1: $3750 \div 25$

Solution: The number 25 is the divisor and the number 3750 is the dividend.

```
        150
   25)3750
      −25↓
       125
      −125↓
        00
        − 0
         0
```

Multiply $1 \times 25 = 25$
Subtract and bring down the 5.
Multiply $5 \times 25 = 125$
Subtract and bring down the 0.
Multiply and subtract
Remainder is 0.

Fill in the correct answers in the space provided.

1. $3,049+27+6,750$ Answer _____

2. $34,598+670+3,409$ Answer _____

3. $375+6,750+9$ Answer _____

4. $709-698$ Answer _____

5. $3,001-2,999$ Answer _____

6. $3,450-2,789$ Answer _____

7. 329×102 Answer _____

8. 675×39 Answer _____

9. 349×210 Answer _____

10. $1,212\div12$ Answer _____

11. $2,457\div63$ Answer _____

12. $595-35$ Answer _____

13. 207×104 Answer _____

14. $309-254$ Answer _____

15. $275+5,750+8$ Answer _____

16. $2,424\div24$ Answer _____

17. 304×12 Answer _____

18. $607-508$ Answer _____

19. 207×21 Answer _____

20. $305+67+3,508$ Answer _____

21. 209×15 Answer _____

22. $309-81$ Answer _____

23. $425-369$ Answer _____

24. $3,075+67,591+397$ Answer _____

25. $398-279$ Answer _____

2. Words Written as Numerals

Integer

An integer is a whole number which may be positive or negative, not a fraction.

1,000,000's	100,000's	10,000's	1,000's	100's	10's	1's
↓	↓	↓	↓	↓	↓	↓
Millions	Hundred Thousands	Ten Thousands	Thousands	Hundreds	Tens	Ones

Example 1:

Write "Seven million twenty-four thousand sixty-eight" in numerals.

Solution:

1,000,000's	100,000's	10,000's	1,000's	100's	10's	1's
↓	↓	↓	↓	↓	↓	↓
7	0	2	4	0	6	8

Answer: 7,024,068

Fill in the letter of the correct answer in the space provided.

1. Fifty-four thousand three hundred is written:
    (a) 54,030   (b) 540,300   (c) 54,300
    (d) 54,000,300   (e) 54,003   Answer _____

2. Three million and twelve is written:
    (a) 30,000,012   (b) 3,000,120   (c) 30,012
    (d) 300,000,012   (e) 3,000,012   Answer _____

3. Ten million one hundred sixty-three is written:
    (a) 10,000,163   (b) 1,000,163   (c) 100,163
    (d) 100,000,163   (e) 100,000,063   Answer _____

4

4. Twenty-three thousand nine is written:
    (a) 2,309   (b) 23,009   (c) 230,090
    (d) 230,000   (e) 23,000,009   Answer _____

5. Seventy-two million thirty nine is written:
    (a) 720,039   (b) 7,200,039   (c) 72,039
    (d) 72,000,039   (e) 720,039   Answer _____

6. Three hundred thousand two hundred ten is written:
    (a) 300,210   (b) 30,210   (c) 3,000,210
    (d) 3,002,100   (e) 30,000,210   Answer _____

7. Six thousand seventy-five is written:
    (a) 67,500   (b) 6,075   (c) 675
    (d) 675,000   (e) 6,750   Answer _____

8. Twenty nine thousand three is written:
    (a) 2,903   (b) 29,030   (c) 290,003
    (d) 2,930   (e) 29,003   Answer _____

9. Five million and sixty-five is written:
    (a) 500,065   (b) 5,000,065   (c) 50,000,065
    (d) 5,000,650   (e) 565,000   Answer _____

3. Adding and Subtracting Units

Example 1:

  4 hours 54 minutes
+2 hours 12 minutes

Solution:

  6 hours   66 minutes
+1 hour   −60 minutes
  7 hours   6 minutes

| Change minutes to hours |
| 66 minutes = 60 minutes + 6 minutes |
| = 1 hour 6 minutes |

Answer: 7 hours 6 minutes

Example 2: Subtract 6 pounds 11 ounces from 11 pounds 3 ounces.

Solution:

Borrow 1 pound = 16 ounces.

$$
\begin{array}{r}
\overset{10}{\cancel{11}} \text{ pounds } \overset{19}{\cancel{3}} \text{ ounces} \\
-\;\; 6 \text{ pounds } 11 \text{ ounces} \\
\hline
4 \text{ pounds } \;\; 8 \text{ ounces}
\end{array}
$$

Answer: 4 pounds 8 ounces

The following information is provided to help you answer the questions in this section:

1 hour   =   60 minutes
1 minute =   60 seconds
1 pound   =   16 ounces
1 yard   =   3 feet
1 foot   =   12 inches

Fill in the correct answers in the space provided.

1.   6 hours 17 minutes
     +2 hours 49 minutes

     Answer _____

2.   12 feet 9 inches
     + 6 feet 7 inches

     Answer _____

3.   4 hours 9 minutes
     −2 hours 12 minutes

     Answer _____

4.   3 feet 7 inches
     −2 feet 12 inches

     Answer _____

5.   6 hours 7 minutes
     −4 hours 12 minutes

     Answer _____

6.   4 feet 2 inches
     −2 feet 7 inches

     Answer _____

7. A movie started at 7:35 P.M. and ended at 10:15 P.M. How long did the movie last?

     Answer _____

8. A flight left Kennedy at 7:42 P.M. and arrived in Los Angeles at 11:29 P.M. How long did it last?

     Answer _____

9. A meeting began at 9:42 A.M. and ended at
   10:19 A.M. How long did the meeting last?   Answer _____

10. A class began at 8:40 A.M. and ends at
    9:45 A.M. How long did the class last?   Answer _____

4. Average Problems

To find the average, add all the numbers and then divide by the amount of numbers added.

Average

The quotient is found by dividing the sum of a group of numbers by the number of addends.

Example 1:   Find the average of 30, 0, and 60.

Solution:

STEP 1:   Add the three numbers.

$$
\begin{array}{r}
30 \\
0 \\
\underline{60} \\
90
\end{array}
$$

STEP 2:   Divide by 3.

$$\frac{90}{3} = 30$$

Answer: 30

Write the answer in the space provided.

1. Find the average of 70, 40 and 0.   Answer _____

2. Find the average of 80, 60 and 40.   Answer _____

3. Jose scored a 90, 80 and 70 on his mathematics
   tests. What is his average?   Answer _____

4. A salesman sells 12 suits on Monday, 14 suits on Tuesday and 16 suits on Wednesday. What is the average number of suits sold for the three days?   Answer _____

5. Find the average of 12, 39 and 9.   Answer _____

6. Lana scored 70 and 80 on her social studies tests. What grade must she receive on the third test to have an 80 average?   Answer _____

7. Zelma scored 90 and 80 on her mathematics tests. What must her grade be on the next test to have a 90 average?   Answer _____

8. John gained 72 pounds in 6 months. What is his average weight gain per month?   Answer _____

9. Find the average of 30, 40 and 50.   Answer _____

10. A salesman sells 3 dresses on Monday, 12 dresses on Tuesday and 15 dresses on Wednesday. What is the average number of dresses sold for the three days?   Answer _____

5. Reducing Fractions

Fraction: It is a part of a whole. The denominator cannot equal zero.

Numerator: The top number in a fraction which tells how many equal parts of the whole are being considered.

Denominator: The bottom number of a fraction which tells the number of equal parts in a fraction. It cannot be zero.

Improper Fraction: A fraction in which the numerator is equal to or greater than the denominator.

Proper Fraction: A fraction which the numerator is less than the denominator.

To reduce a proper fraction to lowest terms, find the largest number that divides evenly into the numerator and into the denominator.

Example 1:   Reduce $\frac{18}{24}$ to lowest terms.

Solution: $\frac{18}{24}$

Answer: $\frac{3}{4}$

Divide numerator and denominator by six $= \frac{3}{4}$

To reduce an improper fraction (numerator larger than the denominator) to lowest terms, divide the denominator into the numerator and then see if the fraction part of your answer can be further reduced.

Example 2:

$\frac{20}{6}$

Divide the denominator into the numerator $= 3\frac{2}{6}$

$3\frac{2}{6}$ Reduce fraction part of your answer $= 3\frac{1}{3}$

Answer: $3\frac{1}{3}$

Fill in the correct answer in the space provided. Reduce to lowest terms.

1. $\frac{3}{9}$ Answer _____   7. $\frac{24}{36}$ Answer _____

2. $\frac{5}{4}$ Answer _____   8. $\frac{36}{24}$ Answer _____

3. $\frac{24}{32}$ Answer _____   9. $\frac{17}{4}$ Answer _____

4. $\frac{32}{24}$ Answer _____   10. $\frac{14}{4}$ Answer _____

5. $\frac{18}{4}$ Answer _____   11. $\frac{18}{27}$ Answer _____

6. $\frac{4}{18}$ Answer _____   12. $\frac{6}{28}$ Answer _____

6. Addition of Fractions

1. Fractions must have a common denominator.

2. Find the lowest common denominator (if denominators are different) and re-evaluate the numerators to find equivalent fractions.

3. Add numerators and re-write the common denominator. Also add the integral parts of the fractions if necessary.

4. Reduce answer to lowest terms.

Least Common Multiple (LCM):   The least multiple excluding 0 of two or more numbers. For example, the (LCM) of 6 and 10 is 30.

Example 1: $\frac{3}{7} + \frac{2}{7} = \frac{5}{7}$

Solution: Both fractions have the same denominator. Add the numerators.

     Answer: $\frac{5}{7}$ Answer is already in lowest terms.

Example 2: $\frac{3}{5} + \frac{2}{7}$

Solution: Find the (LCD) for 5 and 7.

     STEP 1:

         5, 10, 15, 20, 25, 30, $\boxed{35}$

         7, 14, 21, 28, $\boxed{35}$, 42, 49

     The (LCM) for both fractions = 35.

STEP 2:   Re-evaluate the numerator by dividing the original
denominator into the new denominator and multiply
the numerator by the result of the division.

$$\frac{3}{5} + \frac{2}{7}$$

$$\frac{7 \times 3}{35} + \frac{5 \times 2}{35}$$

$$\frac{21}{35} + \frac{10}{35}$$

$$\frac{31}{35}$$

Answer is in
lowest terms.

Answer: $\frac{31}{35}$

Example 3: $3\frac{2}{7} + 4\frac{5}{7}$

Solution:

$$3\frac{2}{7}$$
$$+ \ 4\frac{5}{7}$$
$$\overline{}$$
$$7\frac{7}{7} = 7 + 1 = 8$$

Add the integers and then the fraction part of
the example. Reduce answer to lowest terms.

Fill in the correct answer in the space provided
(reduce answer to lowest terms).

1. $\frac{2}{7} + \frac{1}{6}$ Answer _____

2. $\frac{3}{4} + \frac{1}{8}$ Answer _____

3. $7 + \frac{1}{9}$ Answer _____

4. $\frac{2}{3} + \frac{3}{4}$ Answer _____

8. $\frac{2}{9} + \frac{1}{5}$ Answer _____

9. $\frac{5}{6} + \frac{1}{12}$ Answer _____

10. $8 + \frac{1}{4}$ Answer _____

11. $\frac{2}{7} + \frac{4}{9}$ Answer _____

11

5. $3\frac{6}{7} + 1\frac{1}{7}$ Answer _____

6. $2\frac{1}{4} + 3\frac{5}{8}$ Answer _____

7. $\frac{6}{7} + \frac{1}{5}$ Answer _____

12. $4\frac{1}{5} + 3\frac{4}{5}$ Answer _____

13. $2\frac{3}{5} + 1\frac{5}{8}$ Answer _____

14. $6\frac{1}{4} + 2\frac{3}{7}$ Answer _____

7. Subtraction of Fractions

1. Fractions must have a common denominator.

2. Find the lowest common denominator (if denominators are different) and re-evaluate the numerators to find equivalent fractions.

3. If the first fraction is smaller than the second fraction, borrow "1" from the integral part in the form of a fraction.

4. Subtract numerators and re-write the common denominator. Also, subtract the integral parts of a fraction if necessary.

5. Reduce answer to lowest terms.

Least Common Multiple (LCM):   The least common multiple excluding 0 of two or more numbers. For example the (LCM) of 6 and 10 is 30.

Example 1: $\frac{6}{7} - \frac{2}{7} = \frac{4}{7}$

Solution: Both fractions have the same denominators. Subtract the numerators.

Answer: $\frac{4}{7}$

Example 2: $\frac{3}{7} - \frac{1}{6}$

Solution: STEP 1

Find the (LCD) for 7 and 6.

7, 14, 21, 28, 35, $\boxed{42,}$ 49

6, 12, 18, 24, 30, 36, $\boxed{42}$

The (LCM) for both fractions = 42.

STEP 2:   Re-evaluate the numerator by dividing the original
denominator into the new denominator and
multiply the numerator by the result of
the division.

$$\frac{3}{7} - \frac{1}{6}$$

$$\frac{6 \times 3}{42} - \frac{7 \times 1}{42}$$

$$\frac{18}{42} - \frac{7}{42}$$

$$\frac{11}{42}$$

Answer: $\frac{11}{42}$ $\boxed{\text{Answer is in lowest terms.}}$

Example 3: $6\frac{1}{2} - 3\frac{3}{4}$

STEP 1:   Find the (LCD) = 4

$$6\frac{1}{2} = 6\frac{2}{4} \qquad\qquad\qquad 5\frac{2}{4} + \frac{4}{4} = 5\frac{6}{4}$$
$$-3\frac{3}{4} = -3\frac{3}{4} \qquad\qquad\qquad -3\frac{3}{4}$$
$$\qquad\qquad\qquad\qquad\qquad\qquad 2\frac{3}{4}$$

$\boxed{\text{Since the top fraction is smaller than the bottom fraction, borrow 1 from the 6 in terms of the fraction } \frac{4}{4}.}$

13

Answer: $2\frac{3}{4}$

Fill in the correct answer in the space provided. (reduce answer to lowest terms).

1. $\frac{3}{8} - \frac{1}{8}$ Answer _____

2. $\frac{3}{4} - \frac{1}{8}$ Answer _____

3. $3\frac{2}{7} - 1\frac{1}{6}$ Answer _____

4. $5\frac{4}{9} - 2\frac{1}{4}$ Answer _____

5. $6 - \frac{7}{8}$ Answer _____

6. $\frac{5}{9} - \frac{1}{3}$ Answer _____

7. $3 - \frac{1}{5}$ Answer _____

8. $\frac{2}{5} - \frac{1}{4}$ Answer _____

9. $\frac{3}{4} - \frac{1}{4}$ Answer _____

10. $\frac{2}{7} - \frac{2}{6}$ Answer _____

11. $5\frac{2}{9} - 3\frac{4}{9}$ Answer _____

12. $6\frac{1}{8} - 3\frac{2}{7}$ Answer _____

13. $9 - \frac{2}{3}$ Answer _____

14. $\frac{4}{5} - \frac{1}{3}$ Answer _____

15. $5\frac{1}{4} - 3\frac{2}{7}$ Answer _____

16. $6\frac{1}{8} - 4\frac{5}{7}$ Answer _____

8. Multiplication of Fractions

1. Change improper fractions to proper fractions.

2. Cancel by reducing the numerators and denominators by the same factor.

3. Multiply the numerators and the denominators.

4. Reduce answer to lowest terms.

Factors: Two or more numbers when multiplied give you a product.

Example 1: $\frac{2}{5} \times \frac{1}{7}$

Solution: Numerator and denominator have no common factors. Multiply numerators and denominators. Reduce answer to lowest terms.

$$\frac{2}{5} \times \frac{1}{7}$$

$$\frac{2}{35}$$

Answer: $\frac{2}{35}$

Example 2: $\frac{7}{10} \times \frac{5}{14}$

Solution: Divide 7 and 14 by 7. Divide 5 and 10 by 5.

$$\frac{\overset{1}{\cancel{7}}}{\underset{2}{\cancel{10}}} \times \frac{\overset{1}{\cancel{5}}}{\underset{2}{\cancel{14}}} = \frac{1 \times 1}{2 \times 2} = \frac{1}{4}$$

Example 3: $3\frac{2}{5} \times 3\frac{1}{3}$

Solution:

$$3\frac{2}{5} = 3 + \frac{2}{5} = \frac{15}{5} + \frac{2}{5} = \frac{17}{5}$$

$$3\frac{1}{3} = 3 + \frac{1}{3} = \frac{9}{3} + \frac{1}{3} = \frac{10}{3}$$

$$3\frac{2}{5} \times 3\frac{1}{3}$$

Solution:

1. $\dfrac{17}{\underset{1}{\cancel{5}}} \times \dfrac{\overset{2}{\cancel{10}}}{3}$

2. $\dfrac{34}{3}$

3. $11\frac{1}{3}$

1. Reduce within the problem one numerator with one denominator.

2. Multiply numerators and denominators.

3. Reduce answer to lowest terms. Divide denominator into the numerator (improper fraction). Look at fraction part of your answer to see if it can be further reduced.

15

Fill in the correct answer in the space provided. Reduce the answers to lowest terms.

1. $\frac{2}{7} \times \frac{1}{6}$ Answer _____

9. $\frac{3}{4} \times \frac{8}{9}$ Answer _____

2. $\frac{3}{4} \times \frac{1}{9}$ Answer _____

10. $\frac{2}{7} \times \frac{1}{4}$ Answer _____

3. $6 \times \frac{1}{7}$ Answer _____

11. $2 \times \frac{1}{8}$ Answer _____

4. $\frac{2}{3} \times 9$ Answer _____

12. $\frac{3}{5} \times 25$ Answer _____

5. $\frac{1}{3} \times 6$ Answer _____

13. $3\frac{1}{2} \times 4$ Answer _____

6. $3\frac{2}{5} \times 3\frac{4}{7}$ Answer _____

14. $4\frac{1}{5} \times 2\frac{7}{9}$ Answer _____

7. $2\frac{1}{4} \times 1\frac{5}{9}$ Answer _____

15. $2\frac{1}{7} \times \frac{7}{15}$ Answer _____

8. $\frac{2}{3} \times \frac{1}{5} \times \frac{9}{14}$ Answer _____

16. $\frac{2}{9} \times \frac{5}{6} \times \frac{18}{25}$ Answer _____

9. Division of Fractions

1. Change improper fractions to proper fractions.

2. Invert (turn upside down) the fraction after the division sign.

3. Change the division sign to multiplication.

4. Follow the rules for multiplication. Cancel and then multiply numerators and denominators.

5. Reduce answer to lowest terms.

Invert: Turn upside down a fraction.

Reciprocal: It occurs when the numerator and the denominator of a fraction is interchanged.

Example 1: $\frac{2}{5} \div \frac{7}{9}$

Solution: $\frac{2}{5} \div \frac{7}{9} = \frac{2}{5} \times \frac{9}{7}$

$$= \frac{18}{35}$$

| Divide fractions. Invert the fraction after the division sign called the divisor and then follow rules for multiplication. |

Answer: $\frac{18}{35}$

Example 2: $\frac{7}{24} \div \frac{14}{23}$

Solution: $\dfrac{7}{24} \div \dfrac{14}{23} = \dfrac{\overset{1}{\cancel{7}}}{24} \times \dfrac{23}{\underset{2}{\cancel{14}}}$

Answer: $\frac{23}{48}$

Example 3: $6 \div 3\frac{1}{7}$

Solution: $6 \div 3\frac{1}{7}$

| Improper Fraction: To reduce, divide the denominator into the numerator and see if the fraction part of your answer can be further reduced. |

$$\frac{6}{1} \div \frac{22}{7}$$

$$\dfrac{\overset{3}{\cancel{6}}}{1} \times \dfrac{7}{\underset{11}{\cancel{22}}}$$

$$\frac{21}{11}$$

$$1\frac{10}{11}$$

Answer: $1\frac{10}{11}$

Fill in the correct answer in the space provided. Reduce answers to lowest terms.

1. $\frac{2}{5} \div \frac{4}{5}$ Answer _____

2. $\frac{3}{4} \div \frac{6}{7}$ Answer _____

3. $6 \div \frac{1}{8}$ Answer _____

4. $\frac{2}{7} \div 4$ Answer _____

5. $\frac{2}{3} \div \frac{5}{9}$ Answer _____

6. $1\frac{2}{3} \div 5$ Answer _____

7. $2\frac{1}{4} \div 3\frac{3}{8}$ Answer _____

8. $4\frac{2}{5} \div 1\frac{1}{10}$ Answer _____

9. $\frac{2}{9} \div \frac{2}{27}$ Answer _____

10. $\frac{2}{9} \div \frac{1}{9}$ Answer _____

11. $3 \div \frac{1}{7}$ Answer _____

12. $\frac{4}{9} \div 8$ Answer _____

13. $\frac{2}{9} \div \frac{1}{6}$ Answer _____

14. $3\frac{2}{5} \div \frac{1}{10}$ Answer _____

15. $2\frac{3}{8} \div \frac{9}{16}$ Answer _____

16. $2\frac{1}{5} \div 3\frac{3}{10}$ Answer _____

10. Comparing Size of Fractions

1. Find the equivalent fraction if denominators are different.

2. Compare the numerators. The fraction with the largest numerator is the largest. The fraction with the smallest numerator is the smallest.

3. Sometimes it is easier to cross multiply and to eliminate one of the fractions. Continue this process until the *largest* or *smallest* fraction is found.

    Example 1:   Which of the fractions is the smallest?

      (a) $\frac{2}{15}$ (b) $\frac{7}{15}$ (c) $\frac{11}{15}$ (d) $\frac{13}{15}$ (e) $\frac{1}{15}$

Solution: Each fraction has a denominator of 15. The smallest fraction is the one
              with the smallest numerator.

Answer: $\frac{1}{15}$ Choice (e)

Example 2:   Which is smaller $\frac{2}{7}$ or $\frac{1}{6}$

Solution: Write equivalent fractions.
              Use a denominator of 42.

              $\frac{2}{7} = \frac{6 \times 2}{6 \times 7} = \frac{12}{42}$

              $\frac{1}{6} = \frac{7 \times 1}{7 \times 6} = \frac{7}{42}$

              ┌───┐
              │ Fractions with the same denominator—the smallest │
              │ fraction—smallest numerator. │
              └───┘

Notice: You can also obtain answer by "cross multiplication."

              $\overset{12}{\underset{7}{\frac{2}{7}}} \quad \overset{7}{\underset{6}{\frac{1}{6}}}$

Answer: $\frac{1}{6}$

Example 3:   Which fraction is the smallest?

              (a) $\frac{2}{7}$ (b) $\frac{1}{8}$ (c) $\frac{4}{9}$ (d) $\frac{3}{16}$

Solution: Use the process of elimination by cross multiplication

STEP 1:
              $\overset{16}{\underset{7}{\frac{2}{7}}} \quad \overset{7}{\underset{8}{\frac{1}{8}}}$

              ┌─────────────────────────────────────┐
              │ Cross multiply smallest: $\frac{1}{8}$ │
              └─────────────────────────────────────┘

STEP 2:

$$\overset{9}{1} \quad \overset{32}{4}$$
$$8 \quad 9$$

> Keep the smallest and bring down the next fraction. Cross multiply.

STEP 3:

$$\overset{16}{1} \quad \overset{24}{3}$$
$$8 \quad 16$$

> Keep the smallest and bring down the next fraction. Cross multiply.

Answer: $\frac{1}{8}$

COMPARING FRACTIONS (LARGEST):

Example 4: Which of the fractions is the largest?
(a) $\frac{5}{18}$ (b) $\frac{7}{18}$ (c) $\frac{11}{18}$ (d) $\frac{1}{18}$ (e) $\frac{17}{18}$

Solution: Each fraction has a denominator of 18.
The largest fraction is the one with the largest numerator

Answer: $\frac{17}{18}$ Choice (e)

Example 5: Which is larger $\frac{4}{9}$ or $\frac{1}{8}$?

Solution: Write equivalent fractions.
Use a denominator of 72.

$$\frac{4}{9} = \frac{8 \times 4}{8 \times 9} = \frac{32}{72}$$

$$\frac{1}{8} = \frac{9 \times 1}{9 \times 8} = \frac{9}{72}$$

> Fractions with the same denominator—the largest fraction—the largest numerator.

Notice: You can also obtain answer by "cross multiplication".

Answer: $\frac{4}{9}$

Example 6:   Which fraction is the largest?
(a) $\frac{4}{9}$ (b) $\frac{2}{17}$ (c) $\frac{4}{11}$ (d) $\frac{2}{7}$

Solution: Use the process of elimination by cross multiplication.

STEP 1: $\overset{68}{\underset{9}{4}}$ $\overset{18}{\underset{17}{2}}$ | Cross multiply largest: $\frac{4}{9}$ |

STEP 2: $\overset{44}{\underset{9}{4}}$ $\overset{36}{\underset{11}{4}}$ | Keep the largest and bring down the next fraction. Cross multiply. |

STEP 3: $\overset{28}{\underset{9}{4}}$ $\overset{18}{\underset{7}{2}}$ | Keep the largest and bring down the next fraction. Cross multiply. |

Answer: $\frac{4}{9}$

A. In each exercise, which fraction is the *smallest*? Write the letter of the correct answer in the space provided.

1. (a) $\frac{2}{7}$ (b) $\frac{1}{6}$ (c) $\frac{3}{13}$ (d) $\frac{2}{9}$ (e) $\frac{2}{13}$ Answer _____

2. (a) $\frac{1}{10}$ (b) $\frac{2}{5}$ (c) $\frac{2}{19}$ (d) $\frac{3}{5}$ (e) $\frac{6}{7}$ Answer _____

3. (a) $\frac{1}{5}$ (b) $\frac{3}{7}$ (c) $\frac{2}{15}$ (d) $\frac{1}{6}$ (e) $\frac{2}{9}$ Answer _____

4. (a) $\frac{2}{7}$ (b) $\frac{1}{8}$ (c) $\frac{3}{17}$ (d) $\frac{1}{9}$ (e) $\frac{2}{15}$ Answer _____

5. (a) $\frac{3}{8}$ (b) $\frac{2}{7}$ (c) $\frac{1}{5}$ (d) $\frac{2}{11}$ (e) $\frac{1}{8}$ Answer _____

B. In each exercise, which fraction is the *largest*? Write the letter of the correct answer in the space provided.

1. (a) $\frac{2}{9}$ (b) $\frac{3}{7}$ (c) $\frac{3}{5}$ (d) $\frac{2}{7}$ (e) $\frac{1}{4}$ Answer _____

2. (a) $\frac{1}{5}$ (b) $\frac{2}{7}$ (c) $\frac{3}{15}$ (d) $\frac{2}{17}$ (e) $\frac{1}{3}$ Answer _____

3. (a) $\frac{4}{7}$ (b) $\frac{2}{5}$ (c) $\frac{3}{17}$ (d) $\frac{7}{15}$ (e) $\frac{2}{9}$ Answer _____

4. (a) $\frac{1}{5}$ (b) $\frac{2}{7}$ (c) $\frac{3}{8}$ (d) $\frac{1}{6}$ (e) $\frac{2}{5}$ Answer _____

5. (a) $\frac{4}{9}$ (b) $\frac{3}{8}$ (c) $\frac{1}{8}$ (d) $\frac{1}{9}$ (e) $\frac{2}{7}$ Answer _____

6. (a) $\frac{3}{9}$ (b) $\frac{2}{7}$ (c) $\frac{3}{5}$ (d) $\frac{2}{5}$ (e) $\frac{1}{7}$ Answer _____

7. (a) $\frac{1}{2}$ (b) $\frac{3}{4}$ (c) $\frac{2}{7}$ (d) $\frac{1}{8}$ (e) $\frac{2}{9}$ Answer _____

8. (a) $\frac{3}{7}$ (b) $\frac{2}{5}$ (c) $\frac{3}{17}$ (d) $\frac{4}{19}$ (e) $\frac{1}{5}$ Answer _____

9. (a) $\frac{1}{2}$ (b) $\frac{2}{7}$ (c) $\frac{3}{19}$ (d) $\frac{7}{20}$ (e) $\frac{4}{5}$ Answer _____

11. Comparing size of Decimals

Decimal: A numeral that uses place value digits and a decimal point to write numbers that show tenths, hundredths, thousandths, etc.

Tenth: Being one of ten equal parts (one tenth $= \frac{1}{10}$ or .1).

Hundredth: Being one of a hundred equal parts (one hundredth $= \frac{1}{100} = .01$).

Thousandth: Being one of a thousand equal parts (one thousandth $= \frac{1}{1000} = .001$).

PLACE VALUE:

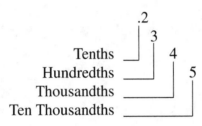

To Compare the Size of Decimals

1. First compare the integral parts.

2. If they are the same, compare the first number after the decimal point (tenths digits).

3. If the tenths digits are the same, then compare the second number after the decimal point (the hundredths digits).

4. Continue this process until the correct answer is found.

   Example 1:   Which is the smallest number?
           (a) 3.04   (b) .042   (c) .031   (d) .03

Solution: In b, c, and d, the integral part is 0. Thus, (a) is eliminated. Now, look at the tenths digit in (b), (c) and (d). The tenths digit is zero for all of them. Look at the hundredths digit. The smallest is (c) and (d) for the hundredths digit. Finally, look at the thousandths digit for (c) and (d). Since (d) has no number in the thousandths digit, it is 0. Thus, the smallest is (d).

Answer: (d)

Example 2: Which is the largest number?

(a) .038   (b) .067   (c) .793   (d) .749

Solution: Look at the tenths digit. The largest is (c) and (d).
Now, look at the hundredths digit. The largest is (c)

Answer: (c)

A. Which number is the *smallest*? Write the letter of
the correct answer in the space provided.

1. (a) .31   (b) .34   (c) .36
    (d) .39   (e) .30   Answer _____

2. (a) .62   (b) .6   (c) .65
    (d) .69   (e) .605   Answer _____

3. (a) .456   (b) .451   (c) .452
    (d) .457   (e) .458   Answer _____

4. (a) 2.75   (b) 2.89   (c) 2.761
    (d) 2.758   (e) 2.77   Answer _____

5. (a) .08   (b) .008   (c) .0008
    (d) .8   (e) .00008   Answer _____

6. (a) 6.75   (b) 6.175   (c) 6.075
    (d) 6.275   (e) 6.891   Answer _____

B. Which number is the *largest*? Write the letter of the correct answer in the space provided.

1. (a) .84   (b) .89   (c) .82   (d) .87   (e) .86   Answer _____

2. (a) .31   (b) .34   (c) .3   (d) .39   (e) .37   Answer _____

3. (a) 6.1   (b) 6.07   (c) 6.21   (d) 6.04   (e) 6.39   Answer _____

4. (a) 3.7   (b) 3.07   (c) 3.75   (d) 3.72   (e) 3.79   Answer _____

5. (a) 5.09   (b) .059   (c) 5.90   (d) 5.009   (e) 5.097   Answer _____

6. (a) .88   (b) .0881   (c) .0808   (d) .0888   (e) .0880   Answer _____

12. Addition and Subtraction of Decimals

1. Line up decimal digits.
2. Add or subtract in appropriate columns.
3. Carry down the decimal point.

Example 1   Add 12.75+19.02+6.71

Solution:

$$
\begin{array}{r}
12.75 \\
19.02 \\
\underline{6.71} \\
38.48
\end{array}
$$

Answer: 38.48

Example 2:   Subtract 39.2−6.97

STEP 1:   Line up decimal points. Convert 39.2 to 39.20
and then proceed to subtract.

$$
\begin{array}{r}
39.20 \\
-\ \ 6.97 \\
\hline
32.23
\end{array}
$$

Answer: 32.23

Fill in the correct answer in the space provided.

A. Addition

1. .271 + 2.71 + 27   Answer _____

2. 6.89 + 35. 02 + .086   Answer _____

3. 16.02 + 1.62 + .162   Answer _____

4. 3.7 + 6.09 + .075   Answer _____

5. 6.9 + 12.7 + 3.85   Answer _____

6. 7.5 + .75 + .075   Answer _____

7. .089 + 89 + 8.9   Answer _____

8. 6.75 + 7.2 + 13.89   Answer _____

9. 7.2 + .72 + 72   Answer _____

10. 6.89 + 68.9 + .075 + 3.8   Answer _____

B. Subtraction

1. 37.2 − 12.1   Answer _____

2. 16.89 − 7.54   Answer _____

3. 3 − .078   Answer _____

4. 16.2 − 3.45   Answer _____

5. 6.78 − 2.89   Answer _____

6. 72 − .72   Answer _____

7. 17.2 − 15.9   Answer _____

8. 6 − .089   Answer _____

9. 424.75 − 189.53   Answer _____

10. 678.2 − 129.87   Answer _____

13. Multiplication of Decimals

1. Multiply as an ordinary problem and obtain the product.

2. Determine where the decimal point belongs by counting the total number of decimal digits after the decimal point in both numbers.

3. Place the decimal point in the product by counting from right to left. If there are not enough digits, then add the appropriate amount of zeros.

Example 1:   Multiply .58 × 7.4

Solution:

$$
\begin{array}{rl}
.58 & \text{2 Decimal Digits} \\
\times\ 7.4 & \underline{+\ \text{1 Decimal Digit}} \\
\hline
232 & \\
\underline{406\ \ } & \\
4.292 & \text{3 Decimal Digits}
\end{array}
$$

Example 2:   Find the cost of 12 pints of cole slaw at $1.39 per pint.

Solution: Each pint of cole slaw cost $1.39. To find the total cost, you multiply the number of pints by the cost of each pint.

$$
\begin{array}{rl}
\$1.39 & \text{2 Decimal Digits} \\
\times\ 12 & \underline{+0\ \text{Decimal Digits}} \\
\hline
278 & \\
\underline{139\ \ } & \\
\$16.68 & \text{2 Decimal Digits}
\end{array}
$$

A. Write the answer in the space provided.

1. .4 × .5   Answer _____   7. .16 × .02   Answer _____

2. .39 × .6   Answer _____   8. 3.2 × 4.5   Answer _____

3. 12 × .08   Answer _____   9. .07 × 18   Answer _____

4. 17 × .001   Answer _____   10. 12 × .0004   Answer _____

28

5. $1.62 \times .05$ Answer _____   11. 16.2×5.1 Answer _____

6. 507×4.16 Answer _____   12. 7.89×653 Answer _____

B. Write the answer in the space provided.

1. How much does three quarts of orange juice cost
    at $.98 a quart?   Answer _____

2. Ice cream cones cost $1.10 each. What is the cost
    of two dozen ice cream cones?   Answer _____

3. Pears sell for $.79 a pound. What is the cost of five
    pounds of pears?   Answer _____

4. Jannie earns $7.35 per hour where she works. How
    much does she earn for twenty-eight hours of work?   Answer _____

5. Coffee sells for $3.75 a pound. What is the cost of
    twelve pounds of coffee?   Answer _____

14. Division of Decimals

1. The number after the division sign is called the divisor. The other number is called the
    dividend. The answer to a division problem is called the quotient.

2. Move the decimal point as many places as necessary to make the divisor an integer.

3. Move the decimal point in the dividend the same amount of places. If there are not
    enough places, the appropriate number of zeros must be added.

4. Proceed to divide. Write the decimal point in the correct place in the quotient.

Example 1: $12.5 \div .05$

Solution: Divisor $\overline{\big|\,\text{Dividend}}^{\text{Quotient}}$

> Move decimal point in divisor two
> places to the right. Then, move
> decimal point in dividend two
> places to the right and add a 0.
> Then, proceed to divide.

$$.05\overline{)12.5}$$

$$\begin{array}{r} 250 \\ 5\overline{)1250} \\ -10 \\ \hline 25 \\ -25 \\ \hline 00 \\ -0 \\ \hline 0 \end{array}$$

Answer: 250

Example 2: Pencils cost $.12 each. How many can be purchased for $24.00?

Solution:

$$.12\overline{)24.00}$$

$$\begin{array}{r} 200 \\ 12\overline{)2400} \end{array}$$

> Move decimal in divisor and in the dividend two places to the right and proceed to divide.

Answer: 200

A. Fill in the correct answer in the space provided.

1. $36 \div .6$ Answer _____
2. $72 \div .03$ Answer _____
3. $140 \div .7$ Answer _____
4. $.49 \div 7$ Answer _____
5. $12.5 \div 2.5$ Answer _____
6. $720 \div .06$ Answer _____
7. $.84 \div 6$ Answer _____
8. $14.4 \div 1.2$ Answer _____
9. $7.5 \div .15$ Answer _____
10. $12.12 \div 12$ Answer _____
11. $160 \div .8$ Answer _____
12. $.42 \div .7$ Answer _____

B. Write the answer in the space provided.

1. Pens cost $.75 each. How many pens can be purchased for $30.00?   Answer _____

2. Apples cost $.25 each. How many can be purchased for $6.00?   Answer _____

3. Nuts sell for $3.60 a pound. How many pounds can be purchased for $72.00?   Answer _____

4. Zelma pays $23.20 for a tank of gasoline. If gasoline sells for $1.45 a gallon, how many gallons does she buy?   Answer _____

15. Conversions, Decimals, Fractions and Percents

1. To change a fraction to a decimal, divide the numerator into the denominator. Add a decimal point and at least two zeros

   Example 1:   Convert $\frac{3}{8}$ to a decimal.

   Solution: Divide the numerator, 3, by the denominator, 8. Add the necessary amount of 0's to the right of the decimal point.

   $$\frac{N}{D}$$

   $$D\overline{)N}$$

   $$\begin{array}{r} .375 \\ 8\overline{)3.000} \\ \underline{24} \\ 60 \\ \underline{56} \\ 40 \\ \underline{40} \\ 0 \end{array}$$

   Answer: .375

A. Change each fraction to a decimal. Fill in the correct answer in the space provided.

1. $\frac{6}{25}$ Answer _____   5. $\frac{9}{2}$ Answer _____

2. $\frac{11}{17}$ Answer _____   6. $\frac{7}{20}$ Answer _____

3. $\frac{9}{200}$ Answer _____   7. $\frac{9}{25}$ Answer _____

4. $\frac{3}{100}$ Answer _____   8. $\frac{7}{4}$ Answer _____

To Convert a Decimal to a Fraction:

Example 1:   Change .08 to a fraction in lowest terms.

Solution: $.08 = \frac{8}{100}$ $\boxed{\text{Divide both the numerator and the denominator by 4.}}$

              $\frac{2}{25}$

Answer: $\frac{2}{25}$

Decimal to Fraction

1. Re-write the decimal as a fraction and reduce answer to lowest terms.

B. Change each decimal to a fraction and reduce answer to lowest terms. Fill in the correct answer in the space provided.

1. .35   Answer _____   6. .045   Answer _____

2. .02   Answer _____   7. .08   Answer _____

3. .025   Answer _____   8. 6.4   Answer _____

4. 1.2   Answer _____   9. .089   Answer _____

5. 4.35   Answer _____   10. .725   Answer _____

Percent to Fraction

1. Cross out the percent symbol.

2. Create a fraction by using 100 as the denominator.

3. Reduce answer to lowest terms.

To Convert a Percent to a Fraction:

Example 1:   Express 65% as a fraction. Reduce answer to lowest terms.

Solution: $65\% = \frac{65}{100} = \frac{13}{20}$

Answer: $\frac{13}{20}$

> Divide both the numerator and the denominator by 5.

C. Change each percent to a fraction. Reduce answer to lowest terms. Fill in the correct answer in the space provided.

1. 35%   Answer _____   6. 70%   Answer _____

2. 8%   Answer _____   7. 79%   Answer _____

3. 19%   Answer _____   8. 245%   Answer _____

4. 24%   Answer _____   9. 36%   Answer _____

5. 150%   Answer _____   10. 4%   Answer _____

Convert a Percent to a Decimal:

1. Place decimal point to the left of the percent symbol, if none is present.

2. Cross out the percent symbol.

3. Move the decimal point two digits to the left.

    Example 1:   Convert 137% to a decimal.

    Solution: 137% = 1.37

        Answer: 1.37

    | Decimal point is moved two places to the left and the percent sign is eliminated. |

D. Change each percent to a decimal. Fill in the correct answer in the space provided.

1. 39%   Answer _____   6. 54%   Answer _____

2. 7%   Answer _____   7. 429%   Answer _____

3. 125%   Answer _____   8. .7%   Answer _____

4. .8%   Answer _____   9. 4%   Answer _____

5. 6%   Answer _____   10. 16%   Answer _____

16. Rounding Off to Nearest Hundredth or Thousandth

1. To round to nearest *hundredth*, the answer must be expressed to *two* decimal digits (two numbers after the decimal point). Look at the third decimal digit (third number after the decimal point). If the third digit is 5 or more, the second decimal digit must be made 1 digit larger (Example: .576=.58 or .574=.57).

2. To round to the nearest *thousandth*, the answer must be expressed to three decimal digits (three numbers after the decimal point). Look at the fourth decimal digit (fourth number after the decimal point). If the fourth digit is 5 or more, the third digit must be made 1 digit larger (Example: .5892 = .589, .5876 = .588).

Example 1:   Convert $\frac{3}{19}$ to a decimal rounded to the nearest hundredth.

Solution: Divide numerator by denominator and add three 0's

```
      .157
  19)3.000
    −19↓
     110
     −95↓
      150
     −133
       17
```

Answer is to be expressed as hundredths (two digits to the right of the decimal point). Since the third digit to the right of the decimal is 5 or greater than 5, then increase the second decimal digit from 5 to 6.

Answer: .16

Example 2:   Convert $\frac{2}{9}$ to a decimal rounded to the nearest thousandth.

Solution: Divide numerator by the denominator and add four 0's.

```
     .2222
  9)2.0000
   −18↓
    20
   −18↓
   −20
    18↓
    20
   −18
    2
```

Answer is to be expressed as thousandths (three digits to the right of the decimal point). Since the fourth digit to the right of the decimal point is not 5 nor greater than 5, the third digit remains 2.

Answer: .222

A. Change each fraction to a decimal rounded to the nearest *hundredth*. Fill in the correct answer in the space provided.

1. $\frac{3}{8}$ Answer _____   5. $\frac{2}{7}$ Answer _____

2. $\frac{2}{9}$ Answer _____   6. $\frac{3}{17}$ Answer _____

3. $\frac{6}{7}$ Answer _____   7. $\frac{8}{9}$ Answer _____

4. $\frac{4}{19}$ Answer _____   8. $\frac{5}{6}$ Answer _____

B. Change each fraction to a decimal rounded to the nearest *thousandth*. Fill in the correct answer in the space provided.

1. $\frac{3}{11}$ Answer _____   5. $\frac{8}{17}$ Answer _____

2. $\frac{10}{13}$ Answer _____   6. $\frac{11}{19}$ Answer _____

3. $\frac{2}{7}$ Answer _____   7. $\frac{4}{7}$ Answer _____

4. $\frac{3}{8}$ Answer _____   8. $\frac{7}{9}$ Answer _____

17. Verbal Problems in Cost and Profit

Example 1:   Candy cost $.40 a bar. How many bars can be purchased for $24?

Solution: Divide the total cost $24 by the cost per bar- $.40.

$24 ÷ $.40

$$.40\overline{)24}$$

$$\frac{60}{40\overline{)2400}}$$

Move decimal in the divisor two places to the right. Add a decimal and two 0's in the dividend and move it two places to the right.

Answer: 60 bars.

Example 2:   18 oranges cost $.15 each. What is the total cost?

35

Solution: Multiply the cost of one orange $.15 by the number of oranges to find the total cost.

$$
\begin{array}{rl}
\$\ .15 & \text{2 Decimal Digits} \\
\times 18 & \underline{\text{0 Decimal Digits}} \\
120 & \\
\underline{\ 15\ } & \\
\$2.70 & \text{2 Decimal Digits}
\end{array}
$$

Answer: $2.70.

Profit = Total Sale − Cost

1. What is the total cost of 35 pencils at 15¢ each?   Answer _____

2. Find the total cost of 3 pounds of apples at 39¢ per pound and 5 pounds of pears at 59¢ per pounds.   Answer _____

3 A piano teacher charges $15.00 for the first lesson and $12.00 for each additional lesson. What is the cost of 9 lessons?   Answer _____

4. The Johnson Moving Company charges $35.00 for the first hour of work and $27.00 for each additional hour. What is the cost of a moving job that takes six hours?   Answer _____

5. A drama group sells 325 tickets to a play written by Cervantes. Each ticket sells for $5.00. The group spends $475.00 to rent the auditorium and $147.00 in additional expenses. What is the profit?   Answer _____

6. A department store buys 75 coats for $3,000.00. All the coats are sold at $89.00 each. What is the profit?   Answer _____

7. Pedro's Superette buys 15 dozen rolls at 79¢ per dozen. Only 11 dozen of the rolls are sold at $1.39 per dozen. What is the profit?   Answer _____

8. Martin Luther King High School sells 635 tickets to a baseball game at $7.00 per ticket. It cost $1,700.00 to rent the stadium. Other expenses amount to $925.00. What is the profit?   Answer _____

9. Elliot spent $9.35 in the Superette. How much change did he receive from a $20.00 bill?   Answer _____

10. Philip bought two candy bars at 65¢ each. How much change did he receive from a $5.00 bill?   Answer _____

18. Percent Problems

1. Change the percent to a decimal.

2. What is 12% of 72? (Change the percent to a decimal and proceed to multiply; $72 \times .12 = 8.64$).

3. If 30% of a number is 60, find the number. (Change the percent to a decimal and proceed to divide; $60 \div .30 = 200$).

Example 1:   What is 30% of 45?

Solution: Change 30% to a decimal.

$$30\% = .30 = .3$$

and then proceed to multiply.

$$.3 \times 45 = 13.5$$

   Answer: 13.5

Example 2:   If 40% of a number is 80, find the number?

Solutions:

   METHOD 1: Change percent a decimal and proceed to divide.

$$80 \div .40 = 200$$

Answer = 200.

37

METHOD 2: Algebraic solution:

Let n = the unknown number

$40\% = .40 = .4$

$80 = .4 \times n$

Divide both sides by .4.

Answer   200

A. Fill in the letter of the correct answer in the space provided.

1. What is 12% of 30?
    (a) 36   (b) 250   (c) 3.6   (d) 360   (e) 25   Answer _____

2. What is 35% of 70?
    (a) 24.5   (b) 245   (c) 200   (d) 20   (e) 2.45   Answer _____

3. If 30% of a number is 60, what is the number?
    (a) 18   (b) 20   (c) 1.8   (d) 2   (e) 200   Answer _____

4. If 25% of a number is 240, find the number.
    (a) 9.6   (b) 60   (c) 6   (d) 960   (e) 96   Answer _____

5. What is 75% of 60?
    (a) 45   (b) 80   (c) 8   (d) 4.5   (e) 800   Answer _____

6. If 70% of a number is 140, find the number.
    (a) 98   (b) 20   (c) 9.8   (d) 200   (e) 980   Answer _____

7. What is 15% of 30?
    (a) 4.5   (b) 20   (c) 45   (d) 2.0   (e) 450   Answer _____

8. What is 30% of 70?
   (a) 200   (b) 21   (c) 37   (d) 69   (e) 210   Answer _____

19. Verbal Problems in Sales Tax and Percent Increase

1. Change the percent to a decimal.

2. Multiply the original amount by the percent in terms of a decimal.

3. Add on sales tax or increase to the original amount.

   Example 1:   Philip earns $85,000. He receives an 8% increase
                in salary. What is his new salary?

   Solution: STEP 1:   Change 8% to a decimal.

                          8% = .08

                STEP 2:   Multiply salary of $85,000 × .08 to
                          find the increase in salary.

                          $85,000 × .08 = $6,800

                STEP 3:   Original Salary + Increase = New Salary.

                          $85,000 + $6,800 = $91,800.

   Answer: $91,800.

   Example 2:   A coat sells for $125.00. There is a 4%
                sales tax. What is the total price?

   Solution: STEP 1:   Change 4% to a decimal.

                          4% = .04

                STEP 2:   Multiply cost of coat at $125.00 × .04.

                          $125.00 × .04 = $5.00

                STEP 3:   Cost + Sales Tax = Total Cost

                          $125.00 × $5.00 = $130.00

   Answer: $130.00.

39

Fill in the correct answer in the space provided.

1. A coat sells for $90.00. There is an 8% sales tax.
    What is the *total price*?   Answer _____

2. Seth Earns $29,000 a year. He receives a 6% increase
    in salary. What is the *new salary*?   Answer _____

3. Eddie buys Zelma a gold bracelet for $475.00. If there
    is an 8% sales tax, what is the *sales tax*?   Answer _____

4. Bill's original pay check was $375.00. It was increased
    by 12%. How much was the *increase*?   Answer _____

5. The population of New York city increased by 4% last
    year. If the original population was 8,000,000, what
    is its *new population*?   Answer _____

6. Box seat tickets to Mets games was increased by 12%.
    Last season the tickets sold for $15.00 each. What do
    they sell for this season?   Answer _____

7. A shirt sells for $19.50 plus a 7% sales tax. What is
    the total price?   Answer _____

20. Verbal Problems in Discount and Percent Decrease

1. Change the percent to a decimal and then multiply.

2. Subtract the discount or the decrease.

Sale Price = Original Price − Discount

Example 1:   In order to keep his job, Seth took a 4%
             decrease in salary. His original salary was
             $95,000. What is his new salary?

Solution: STEP 1:   Change 4% to a decimal

                       $4\% = .04$

STEP 2: Multiply the salary $95,000 by the percentage decrease in salary of 4%.

$95,000 × .04 = $3,800

STEP 3: New Salary = Original Salary − Decrease
New Salary = $95,000 − $3,800
New Salary = $91,200

Answer: $91,200

Example 2: A coat sold for $125.00. The discount is 20%. What is the final price?

Solution: STEP 1: Change 20% to a decimal.

20% = .20 = .2

STEP 2: Multiply the cost of the coat of $125.00 by the percentage discount of 20%.

$125.00 × .2 = $25.00 Discount

STEP 3: Final Price = Original Price − Discount
Final Price = $125.00 − $25.00
Final Price = $100

Answer: $100.00

Fill in the correct answer in the space provided.

1. Elliot's weekly salary is 425.00. It is reduced by 9%. What is his new weekly salary? Answer _____

2. There is a 20% decrease on all luggage. Zelma bought an attache case that originally sold for $89.00. What is the sale price? Answer _____

3. New York City has 8,000,000 people. It loses 7% of its population. What is the new population? Answer _____

4. Philip earns $25,000 a year. He has 12% of his salary deducted for taxes. What is his yearly take-home pay? Answer _____

5. A dress that sold for $37.95 is reduced by 8%. What is the sale price? Answer _____

41

6. Seth buys a sweater that originally sells for $25.00. It is reduced by 35%. What is the sale price?

    Answer _____

7. Milk sells for $2.40 a gallon. The price is reduced by 30%. What is the sale price?

    Answer _____

8. The population of a city is reduced by 40%. If the original population was 3,300,000, what is its new population?

    Answer _____

21. Verbal Problems in Area, Perimeter and Cost

Formulas

1. Area of Rectangle − length × width (square units).

2. The Perimeter of a closed figure is the sum of the lengths of all its sides.

Perimeter of Rectangle = 2 lengths + 2 widths

Area: It is the number of square units needed to cover the surface to be measured.

Perimeter: The perimeter of a closed figure, such as a rectangle, is the sum of all the sides.

Example 1: Find the area of a rectangle whose length is 18 inches and whose width is 6 inches.

Solution: STEP 1: Write the formula.

$$Area = length \times width$$

$$A = lw$$

STEP 2: Substitute and evaluate.

$$A = 18 \times 6$$

$$A \times 108 \text{ in}^2 \text{ (square inches)}$$

Answer: 108 inches^2

Example 2: What is the cost to carpet a room 15 yards by 8 yards at $12 per square yard?

Solution: STEP 1: Find the area.

$$A = l \times w$$
$$A = 15 \times 8$$
$$A = 120 \text{ square yards}$$

STEP 2: Multiply the number of square yards (120) by the cost per square yard ($12.00).

$$120 \times \$12.00 = \$1,440.00$$

Answer: $1,440.00

Example 3: Find the perimeter of a rectangle whose length is 12 feet and whose width is 8 feet.

Solution: STEP 1: Write the formula.
Perimeter = 2 length + 2 width
$$P = 2l + 2w$$

STEP 2: Substitute and evaluate.

$$P = 2(12) + 2(8)$$
$$P = 24 + 16$$
$$P = 40 \text{ feet}$$

Answer: 40 feet

Example 4: Find the cost to fence in a garden that is 30 feet long and 20 feet wide at a cost of $6.00 per foot for fencing.

Solution: STEP 1: Write the formula.
$$P = 2l + 2w$$

STEP 2: Substitute and evaluate.

$$P = 2(30) + 2(20)$$
$$P = 60 + 40$$
$$P = 100 \text{ feet}$$

STEP 3: Multiply the perimeter 100 ft. by the cost per foot at $6.00.

$$100 \times \$6.00 = \$600.00$$

Answer: $600.00

A. Fill in the correct answer in the space provided.

1. Find the area of a room that is 8 yards by 5 yards.   Answer _____

2. Find the area of a tablecloth that is 65 inches long and 30 inches wide.   Answer _____

3. How much does it cost to carpet a room 9 yards by 8 yards at $7.00 per square yard?   Answer _____

4. Carpeting costs $9.00 per square yard. How much does it cost to carpet a room 7 yards by 5 yards?   Answer _____

5. Find the perimeter of a rectangle whose length is 9 feet and whose width is 6 feet.   Answer _____

6. How much fencing is needed to fence in a swimming pool 35 feet long and 20 feet wide?   Answer _____

7. Fencing cost $6.00 per yard. How much does it cost to fence in a rectangular yard that is 40 yards by 30 yards?   Answer _____

8. How much does it cost to fence in a rectangular garden 18 feet by 9 feet at $7.00 per foot of fencing?   Answer _____

9. Carpeting cost $12.00 per square yard. How much does it cost to carpet a room 9 yards by 8 yards?   Answer _____

10. Find the perimeter of a triangle whose sides are 3, 4 and 5 inches.   Answer _____

22. Arithmetic Examinations and Answers

ARITHMETIC EXAM # 1

A. Write the letter of the correct answer in the space provided

1. Find the average of 30, 40, 0, and 50.
   (a) 40　　　　　(b) 120　　　　　(c) 30
   (d) 20　　　　　(e) 60　　　　　　　　　　　　Answer _____

2. Write three million thirty thousand fifty-six.
   (a) 330,056　　(b) 3,030,056　　(c) 30,056
   (d) 3,030,560　(e) 30,030,056　　　　　　　Answer _____

3. $649 - 56$ equals:
   (a) 593　　　　(b) 705　　　　　(c) 583
   (d) 603　　　　(e) 503　　　　　　　　　　　Answer _____

4. 　11 hours 40 minutes
   　$-$ 6 hours 56 minutes
   (a) 5 hours 16 minutes　　(b) 3 hours 2 minutes
   (c) 4 hours 12 minutes　　(d) 4 hours 44 minutes
   (e) 3 hours 12 minutes　　　　　　　　　　　Answer _____

5. $\frac{3}{7} + \frac{1}{6}$
   (a) $\frac{4}{13}$　　　　(b) $\frac{11}{21}$　　　　(c) $\frac{2}{21}$
   (d) $\frac{3}{42}$　　　　(e) $\frac{25}{42}$　　　　　　　Answer _____

6. $1212 \div 12$

    (a) 11   (b) 101   (c) 202

    (d) 22   (e) 102   Answer _____

7. $3\frac{1}{6} - 1\frac{5}{6}$

    (a) $1\frac{1}{3}$ (b) $1\frac{1}{2}$ (c) $2\frac{1}{3}$

    (d) $2\frac{1}{2}$ (e) $1\frac{2}{3}$ Answer _____

8. $3.08 + 30.8 + 1.2$ equals:

    (a) 35.8   (b) 3.58   (c) 35.08

    (d) 3.508   (e) 350.8   Answer _____

9. $\frac{2}{3} \div \frac{1}{4}$

    (a) $\frac{1}{6}$ (b) $\frac{3}{8}$ (c) $\frac{5}{8}$

    (d) $\frac{2}{3}$ (e) $2\frac{2}{3}$ Answer _____

10. Which of the fractions is the smallest?

    (a) $\frac{2}{7}$ (b) $\frac{1}{6}$ (c) $\frac{4}{9}$

    (d) $\frac{4}{7}$ (e) $\frac{2}{9}$ Answer _____

11. A theater club sells 312 tickets to a play at $6.00 each. The club rents a theater for $450.00 and it has $325.00 in other expenses. How much profit does the club make?
    (a) $1,547   (b) $1,422   (c) $997
    (d) $1,457   (e) $1,097   Answer _____

12. $39.1 - 12.76$ equals:
    (a) 26.34   (b) 16.34   (c) 6.34
    (d) 36.34   (e) 26.46   Answer _____

13. What is 30% of 40?
    (a) 120   (b) 75   (c) 60
    (d) 130   (e) 12   Answer _____

14. $3(-2)^2 + 4(-7)$
    (a) 16   (b) 40   (c) -40
    (d) -16 (e) 160   Answer _____

15. If 60% of a number is 30, find the number.
    (a) 18   (b) 180   (c) 50
    (d) 5   (e) 72   Answer _____

16. Change $\frac{6}{7}$ to a decimal rounded to the nearest hundredth.
    (a) .86   (b) .85   (c) .857
    (d) .83   (e) .81   Answer _____

17. A $20 shirt is reduced by 30%. What is the sale price?
    (a) $6   (b) $14   (c) $12
    (d) $8   (e) $15   Answer _____

18. Which number is the smallest?
    (a) .035   (b) .305   (c) 3.05
    (d) .036   (e) .0305   Answer _____

19. Find the cost to carpet a room 8 yards by 5 yards
    at $6 per square yard.
    (a) $48   (b) $240   (c) $120
    (d) $156   (e) $96   Answer _____

20. 3.4 ÷ .2 equals:
    (a) 17   (b) .17   (c) 1.7
    (d) .017   (e) 170   Answer _____

ANSWERS: ARITHMETIC EXAM # 1

1.	c	11.	e
2.	b	12.	a
3.	a	13.	e
4.	d	14.	d
5.	e	15.	c
6.	b	16.	a
7.	a	17.	b
8.	c	18.	e
9.	e	19.	b
10.	b	20.	a

ARITHMETIC EXAM # 2

B. Write the letter of the correct answer in the space provided.

1. Write two million thirty-seven.
   (a) 20,000,037   (b) 2,000,037   (c) 2,000,307
   (d) 200,037   (e) 200,307   Answer _____

2. 701 − 49 equals:
   (a) 652   (b) 750   (c) 662
   (d) 752   (e) 692   Answer _____

3. $\frac{2}{7} + \frac{1}{4}$

   (a) $\frac{3}{11}$ (b) $\frac{2}{7}$ (c) $\frac{2}{14}$
   (d) $\frac{15}{28}$ (e) $\frac{6}{7}$ Answer _____

4.   9 hours 12 minutes
   −4 hours 16 minutes
   (a) 5 hours 56 minutes   (b) 4 hours 34 minutes
   (c) 4 hours 56 minutes   (d) 5 hours 34 minutes
   (e) 3 hours 12 minutes   Answer _____

5. $8.47 - .59$ equals:
   - (a) 8.88
   - (b) 78.8
   - (c) 64.7
   - (d) 6.08
   - (e) 7.88

   Answer _____

6. Find the average of 30, 40, and 80:
   - (a) 150
   - (b) 50
   - (c) 30
   - (d) 35
   - (e) 55

   Answer _____

7. $1\frac{2}{3} \div 6$
   - (a) 10
   - (b) $\frac{5}{18}$
   - (c) $3\frac{3}{5}$
   - (d) $\frac{2}{9}$
   - (e) $\frac{4}{9}$

   Answer _____

8. $3.7 + .37 + 37$ equals:
   - (a) 4.44
   - (b) 4.107
   - (c) 40.07
   - (d) 4.007
   - (e) 41.07

   Answer _____

9. What is 12% of 72?
   - (a) 8.64
   - (b) 7.92
   - (c) 86.4
   - (d) 79.2
   - (e) 7.64

   Answer _____

10. $6(-1)^2 + 3(-6)$
    - (a) -24
    - (b) -12
    - (c) 24
    - (d) 12
    - (e) -18

    Answer _____

11. If 30% of a number is 30, find the number:
    - (a) 9
    - (b) 10
    - (c) .9
    - (d) 100
    - (e) 90

    Answer _____

12. Which number is the smallest?
    (a) .04   (b) .41   (c) .047
    (d) .004   (e) .0045   Answer _____

13. $2 \div .2$ equals:
    (a) 1   (b) 10   (c) .1
    (d) 100   (e) .01   Answer _____

14. Change $\frac{3}{7}$ to a decimal rounded to the nearest hundredth.

    (a) .42   (b) .428   (c) .43
    (d) .41   (e) .429   Answer _____

15. Express 65% as a fraction:

    (a) $\frac{13}{20}$ (b) $\frac{13}{100}$ (c) $1\frac{7}{20}$
    (d) $\frac{13}{200}$ (e) $3\frac{7}{20}$ Answer _____

16. 20% of the students of City College register for biology.
    If 200 students register for biology, how many students
    are enrolled at City College?
    (a) 40   (b) 400   (c) 10,000
    (d) 100   (e) 1,000   Answer _____

17. If pens sell for \$.15 each, how many can be purchased
    for \$45?
    (a) 300   (b) 62   (c) 3,000
    (d) 620   (e) 30   Answer _____

18. $3.2 \times .001$ equals:
    (a) .3200   (b) .032   (c) .0032
    (d) .32   (e) .00032   Answer _____

19. $3\frac{5}{8} - 1\frac{1}{4}$
    (a) $2\frac{1}{2}$ (b) $3\frac{3}{8}$ (c) $4\frac{7}{8}$
    (d) $2\frac{3}{8}$ (e) $4\frac{1}{2}$ Answer _____

20. $12 \div 1.2$ equals:
    (a) 100   (b) 1000   (c) .1
    (d) .01   (e) 10   Answer _____

ANSWERS: **ARITHMETIC EXAM #2**

1.	b	11.	d
2.	a	12.	d
3.	d	13.	b
4.	c	14.	c
5.	e	15.	a
6.	b	16.	e
7.	b	17.	a
8.	e	18.	c
9.	a	19.	d
10.	b	20.	e

Part II: Algebra — Geometry

23. Rules for Sign Numbers: Addition, Subtraction, Multiplication and Division

The Real Numbers:

A **Set** is a collection of objects. Sets that are parts of other sets are called *subsets*.

Natural Numbers: {1, 2, 3 . . . } These are sets of numbers used for counting.

Whole Numbers: {0, 1, 2, 3 . . . } This is the set of natural numbers with 0 included.

Integers: Integers are whole numbers not fractions.

What are *Real Numbers?* They are a set of all rational numbers and irrational numbers. All these numbers can be represented on a number line.

Rational Numbers: Any number that can be represented as a single fraction with a non-zero divisor.

Irrational Numbers: They are real numbers that are not rational such as: $\sqrt{3}$ *or* $\sqrt{5}$

The Number Line: Is used to represent all positive numbers and all negative numbers.

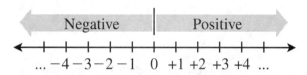

When representing a number on the number line, the distance of that number from 0 is called the absolute value. Thus, the absolute value of $+2$ is 2, and the absolute value of -2 is 2. The absolute value of a number is its distance from 0 on the number line. The absolute value of a number is represented by the symbol ‖.

Example 1:   Find the sum using a number line $(+3) + (+2)$.

Solution: This means begin at 0 and go 3 units to the right (because 3 is positive). Then go 2 more to the right (because 2 is also positive). You should now be at +5, showing that (+3) + (+2) = +5 or 3 + 2 = 5. Using the number line it appears as follows:

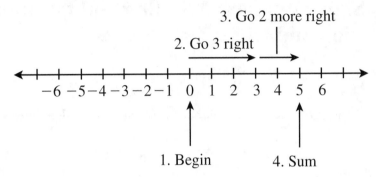

Answer: 5

Addition of Sign Numbers

1. When adding numbers with the same signs, keep the sign and add the numbers.

2. When adding numbers with the opposite signs, keep the sign of the larger number and take the difference.

Example 1:   (−3) + (−5)

Answer: −8

Example 2:   (−6) + (8)

Answer: 2

When adding more than 2 numbers (negative and positive), the following rules apply:

> **Commutative Law of Addition: a + b = b + a**
> **Associative Law of Addition: (a + b) + c = a + (b + c)**

For example,

Commutative Law: 7 + 4 = 4 + 7 = 11
Associative Law: (8 + 1) + 6 = 8 + (1 + 6) = 15

Example 3: (−4) + (7) + (−8) + (3) + (−7) + (4)

HINT: Make columns of all positive and negative numbers.

-4	7
-8	3
-7	4
-19	14

Sum: $-19 + (14) =$
$\qquad -5$

Answer: -5

Fill in the correct answer in the space provided.

1. $(-3) + (-4)$ Answer _____

2. $-2 + (+7)$ Answer _____

3. $(-9) + (12)$ Answer _____

4. $(7) + (-4)$ Answer _____

5. $-3 + (-8)$ Answer _____

6. $6 + (-7)$ Answer _____

7. $-8 + 4 - 7 + 6 - 3 + 8$ Answer _____

8. $-2 - 7 + 6 - 3 + 2 + 1$ Answer _____

Subtraction of Sign Numbers

1. When subtracting numbers, change the sign of the subtrahend (the number after the subtraction sign) to the opposite sign and then follow rules for addition.

    Example 1: $6 - (-5)$

    Solution: The opposite of -5 is 5.
                            $-(-5) = 5$
                         $6 - (-5) = 6 + 5 = 11$

      Answer: 11

Example 2: $-6 - (-5)$

Solution: The opposite of -5 is 5.
                    $-(-5) = 5$
                    $-6 - (-5) = -6 + 5 = -1$

Answer: -1

Fill in the correct answer in the space provided.

1. $(-2) - (-6)$ Answer _____

2. $(-3) - (-7)$ Answer _____

3. $-2 - (-7)$ Answer _____

4. $4 - (7)$ Answer _____

5. $6 - (-3)$ Answer _____

6. $-6 - (-2)$ Answer _____

7. $9 - (-5)$ Answer _____

8. $-8 - (-6)$ Answer _____

9. $-1 - (10)$ Answer _____

10. $-3 - (-4)$ Answer _____

11. $(-2) - (-7)$ Answer _____

12. $3 - (-8)$ Answer _____

Multiplication and Division of Sign Numbers

1. When multiplying or dividing *two* numbers with the *same* signs, the result is positive.

2. When multiplying or dividing *two* numbers with the *opposite* signs, the result is negative.

3. The product of an odd number of negative numbers is negative.

4, The product of an even number of negative numbers is positive.

Multiplication of Sign Numbers:

Numeral: The symbol used to represent numbers.

Variable: A letter that represents a number.

A dot between two numerals or variables indicates multiplication. (2·3 or A·B)

> **a·b = b·a: Commutative Law of Multiplication**
> **(a·b) c = a (b·c): Associative Law of Multiplication**
> **a·1 = 1·a = a: Multiplicative Identity.**
> **   One times any number is that number.**

For Example:

$6 \times 7 = 7 \times 6 = 42$:   Commutative Law of Multiplication

$(3 \times 4) \times 5 = 3 \times (4 \times 5) = 60$:   Associative Law of Multiplication

Example 1: $3(-7)$

Solution: $3(-7) = -21$

> No sign in front of a number indicates it is positive.

Answer: -21

Example 2: $(-3) \; (-4)$

Solution: $(-3) \; (-4) = 12$

Answer: 12

Example 3: $(-2) \; (-7) \; (-1) \; (-5) \; (-6)$

Solution: Multiplication. An odd number of negative signs the answer is negative.

Answer: -420

Example 4: $(-3) \; (-4) \; (-1) \; (-5) \; (-2) \; (-6)$

Solution: Multiplication. An even number of negative signs the answer is positive.

Answer: 720

DIVISION FACTS:

1. $\frac{a}{a} = 1$ A number divided by itself equals 1.

2. $\frac{a}{1} = a$ A number divided by 1 equals the number.

3. $\frac{0}{a} = 0$ Zero divided by a number equals zero.

4. $\frac{a}{0} = \infty$ Undefined. A number divided by zero is undefined — not defined.

Example 1: $\frac{-6}{2}$

Solution: $\frac{-6}{2}$

Numerator and Denominator: Opposite signs in division. The result is negative (no sign by the number 2 indicates it is positive).

-3

Answer: -3

Example 2: $\frac{-9}{-3}$

Solution: $\frac{-9}{-3}$

Numerator and denominator: Same signs in division means the result is positive.

3

Answer: 3

Fill in the correct answer in the space provided.

1. $(3)\ (-4)$ Answer _____

2. $7(-5)$ Answer _____

3. $-6(-2)$ Answer _____

4. $(-3)\ (5)$ Answer _____

5. $(-1)\ (-2)\ (-4)\ (-1)\ (-7)$ Answer _____

6. (-3) (-1) (-4) (-1) (-5) (-1) Answer _____

7. $\frac{-63}{-7}$ Answer _____

8. $\frac{-15}{5}$ Answer _____

9. $\frac{32}{-8}$ Answer _____

10. $\frac{-9}{-3}$ Answer _____

11. $\frac{-14}{2}$ Answer _____

12. $\frac{63}{-7}$ Answer _____

24. Order of Operations

Rules for Order of Operations

1. Perform indicated operations inside the parenthesis.

2. Evaluate the roots and the powers.

3. Multiply or divide from left to right.

4. If necessary, add or subtract.

Squaring a Number: A number multiplied by itself.

Power: It is the number of times a base is used as a factor.

Exponent: It is the number of times a factor is used.

Base: The number that serves as a starting point.

$Base^{Power}$

1. 3^4:   3 Base $-$ 4 power.

2. 6^7:   6 Base $-$ 7 power.

For example:

1. $a^3 = a \cdot a \cdot a$
2. $2^3 = 2 \cdot 2 \cdot 2 = 8$

3. $a^5 = a \cdot a \cdot a \cdot a \cdot a$
4. $3^5 = 3 \cdot 3 \cdot 3 \cdot 3 \cdot 3 = 243$

	A dot between two variables or numerals indicates multiplication.

Example 1: $3(-2 + 4)$

Solution: $3(2)$

6

Answer: 6

	Perform operation within parentheses. Multiplication.

Example 2: $3 - 2 (4 + 5)^2$

Solution: $3 - 2 (9)^2$

$3 - 2 (81)$

$3 - 162$

-159

Answer: -159

Perform operation within parentheses. Powers.
Multiplication (2 numbers opposite signs the result is negative.)
Addition (2 numbers opposite signs keep the sign of the larger number and take the difference.)

Example 3: $4(-5)^2 + 3(-7)$

Solution: $4(25) + 3(-7)$

$100 - 21$

79

Answer: 79

Power
Multiplication (left to right). Two numbers same signs, result is positive. Two numbers opposite signs, the result is negative.
Addition: Signs are different, keep the sign of the larger number and take the difference.

A. Write the correct answer in the space provided.

1. -3^2 Answer _____

2. $(-3)^2$ Answer _____

3. $6(7+1)$ Answer _____

4. $8(4-2)$ Answer _____

5. $-2(8-4)$ Answer _____

6. $6(3-7)$ Answer _____

7. $2-3(7-4)$ Answer _____

8. $6+2(8+4)$ Answer _____

9. $3(8+1)+4$ Answer _____

10. $7(4+3)-3(5+9)$ Answer _____

11. $4(6)^2$ Answer _____

12. $-3(4+5)^2$ Answer _____

13. $-3+(4+5)^2$ Answer _____

14. $6-7(5)^2$ Answer _____

15. $-3(-4)^2$ Answer _____

16. $-6(-3-4)$ Answer _____

17. $3+2(4+5)^2$ Answer _____

18. $(-3)^2+4(-5)$ Answer _____

19. $(-4-5)^2$ Answer _____

20. $(-2)^2-4(-3)^2$ Answer _____

61

25. Interpreting Algebraic Expressions

Variable: A symbol that stands for a number.

Example 1:   Suppose that "b" represents a number.

(a) 2 less than a number

(b) 5 more than a number

(c) 3 times a number

(d) the number divided by 4

Solution: (a) Less than represents $-$, $b-2$

(b) More than represents $+$, $b+5$ or $5+b$

(c) 3 times a number, $3 \cdot b$ or $3b$

(d) The number divided by 4, $b \div 4$ or $\frac{b}{4}$

Example 2:   Find the value in cents of n nickels and d dimes.

Solution: One nickel is valued at 5 cents.
n nickels $= 5 \times n = 5n$

One dime is valued at 10 cents.
d dimes $= 10 \times d = 10d$

n nickels and d dimes
$5n + 10d$

Answer: $5n + 10d$

Interpret each of the following algebraic expressions in terms of "x":
Fill in the correct answer in the space provided.

1. 5 more than a number   Answer _____

2. a number minus 7   Answer _____

3. a number decreased by 6   Answer _____

4. a number plus 5   Answer _____

5. 7 times a number   Answer _____

6. 4 less than a number   Answer _____

7. a number increased by 14   Answer _____

8. 4 more than one half a number   Answer _____

9. 3 more than 4 times a number   Answer _____

10. 6 less than 3 times a number   Answer _____

Interpret each of the following algebraic expressions. Fill in the correct answer in the space provided.

Value of Money:

1. One nickel = 5 cents
   "n" nickels = 5n

2. One dime   = 10 cents
   "d" dimes   = 10d

3. One quarter = 25 cents
   "q" quarters = 25q

1. Find the value in cents of "n" nickels and
   "q" quarters.   Answer _____

2. Find the value in cents of "d" dimes and
   "q" quarters.   Answer _____

3. Find the value in cents of "d" dimes and
   "n" nickels.   Answer _____

4. Find the value in cents of "q" quarters, "n" nickels
   and "d" dimes.   Answer _____

26. Addition and Subtraction of Algebraic Expressions

> **RULES**
> **Addition and Subtraction of Algebraic Expressions**

1. Add or subtract coefficients (numbers in front) of like terms.

63

Like Terms:

    A. Same variables

    B. Same exponents for the same variables.

Example 1:   Which are like terms?

        (a) $6x^2, 3x^2$ (b) $4a^2, 6a^3$

Solution:

    (a) Like Terms: Both terms have the same variable and the same exponent.
                             (a) $6x^2, 3x^2$.

    (b) Unlike Terms: Both terms have the same variable. However, they have different exponents.
                             (b) $4a^2, 6a^3$.

Example 2: $(3a + 4b) + (5a - 7b)$

Solution: Line up like terms and ADD in columns.

$$3a + 4b$$
$$\underline{5a - 7b}$$
$$8a - 3b$$

Answer: $8a - 3b$

Example 3:   Find the sum of $3x^2 - 7x$ and $2 - 5x$.

Solution: Line up like terms.

$$3x^2 - 7x$$
$$\underline{-5x + 2}$$
$$3x^2 - 12x + 2$$

Answer: $3x^2 - 12x + 2$

Example 4:   Subtract $6x^2 - 9$ from $3x^2 - 7$.

Solution: 1) Line up like terms.

$$3x^2 - 7$$
$$\underline{6x^2 - 9}$$

2) Change signs of subtrahend (bottom number to the opposite) and then follow rules for addition.

$$3x^2 - 7$$
$$\underline{-6x^2 + 9}$$
$$-3x^2 + 2$$

Answer: $-3x^2 + 2$

Fill in the correct answer in the space provided.

1. Add $3x + 7y$ and $6x + 8y$ Answer _____

2. Add $3a - 7b$ and $4a - 2b$ Answer _____

3. Add $6x^2 + 5x$ and $8x^3 - 7x$ Answer _____

4. Add $7b - 2$ and $3b + 7$ Answer _____

5. Find the sum of $4x^2 - 7x + 6$ and $-3x + 2$ Answer _____

6. Find the sum of $7y^2 + 3y$ and $3y^2 + 7y - 8$ Answer _____

7. Add $3a^2 - 7a + 6$ and $-7a^2 + 4a - 8$ Answer _____

8. Subtract $-3x + 6$ from $x - 8$ Answer _____

9. Subtract $-7x^2 + 8x$ from $x^2 - 8x$ Answer _____

10. Subtract $-3x^2 - 7x$ from $x - 7$ Answer _____

11. Subtract $7y - 3$ from $-7y - 3$ Answer _____

12. Subtract $-7x^2 + 3x - 7$ from $x^2 - 8x + 9$ Answer _____

13. Subtract $3x - 4$ from $x - 7$ Answer _____

14. Subtract $3x^2 - 7x + 6$ from $-3x^2 + 7x - 6$ Answer _____

15. Simplify $(7x^2 + 3x - 6) - (-x^2 + 6)$ Answer _____

16. Simplify $(-3x^2 + 6) + (3x - 9)$ Answer _____

17. Simplify $(2x^2 - 7x) + (-3x - 6)$ Answer _____

18. Simplify $(-3x^2 + 6x - 9) - (3x^2 + 6x - 9)$ Answer _____

27. Exponent Rule for Multiplication and Monomial Multiplication

Exponent Rule for Multiplication

1. When multiplying numbers with the same base, keep the base and add the exponents. $(A^m)(A^n) = A^{m+n}$

2. If no exponent is written, it is understood to be 1.

Monomial: Is a polynomial with only one term.

Binomial: Is a polynomial with two terms.

Trinomial: Is a polynomial with three terms.

Polynomial: Is an algebraic expression in which all the exponents are whole numbers and in which there is no division by a variable (no variable in the denominator of a fraction).

Example 1: $c^7 \cdot c^2$

Solution: $c^7 \cdot c^2 = c^{7+2} = c^9$

Answer: c^9

Example 2: $a^7 \cdot a$

Solution: Notice $a = a^1$

$a^7 \cdot a = a^7 \cdot a^1 = a^{7+1} = a^8$

Answer: a^8

Fill in the correct answer in the space provided.

1. $(X^2)(X^3)$ Answer _____

2. $(X^2)(Y^3)$ Answer _____

3. $(A^2)(A)$ Answer _____

4. $(A^7)(A^2)$ Answer _____

5. $(2^6)(2^5)$ Answer _____

6. $(A^3)(A)$ Answer _____

7. (Y^2) (Y^7) Answer _____

8. (Y^8) (X) Answer _____

9. (Y^3) (X) Answer _____

10. (Y^2) (Y^7) Answer _____

28. Multiplication of a Monomial by a Monomial

Rules for Multiplication of a Monomial by a Monomial

1. Determine the sign of the product.

2. Multiply all the coefficients (the numbers in front of the unknowns).

3. Multiply the variables by adding exponents, if the bases are the same.

Example 1: $(-6x^2y^7)$ $(3x^4y^2)$

Solution: STEP 1:   Sign of answer $-$

                  STEP 2:   Multiply coefficients (numbers in front of the variables).

                         18

                  STEP 3:   Multiplication: Same base $-$ add exponents.

$$x^2 \cdot x^4 = x^6$$
$$y^7 \cdot y^2 = y^9$$
$$-18x^6y^9$$

Answer: $-18x^6y^9$

Fill in the correct answer in the space provided.

1. $(-3x^2y)$ $(-4xy^2)$ Answer _____

2. $(7xy^2)$ $(-2x^4y^6)$ Answer _____

3. $(7rst)$ $(-2rst)$ Answer _____

4. $(-6x^2y^3)$ $(2x^3y^8)$ Answer _____

5. $(3ab^2c)$ $(-4abc)$ Answer _____

6. $(-4x^3) \ (-8x^2)$ Answer _____

7. $(7xy^2z) \ (-2xyz)$ Answer _____

8. $(3ab^2c) \ (-4a^2bc^2)$ Answer _____

9. $(7rst^2) \ (-2r^2st)$ Answer _____

10. $(-6x^2y^2z) \ (-2x^2y^2z^2)$ Answer _____

29. Multiplication of a Monomial by a Binomial

DISTRIBUTIVE LAW OF MULTIPLICATION:

This law holds for multiplication over addition.
For any three numbers a, b, and c, multiplication
before addition: $a(b + c) = ab + ac$

Example 1: $3x^2 \ (4x^3 - 6x)$

Solution: $3x^2 \ (4x^3 - 6x)$
                 $3x^2{\cdot}4x^3 - 3x^2{\cdot}6x$ Distributive Law

Answer: $12x^5 - 18x^3$

Example 2: $3a \ (a^2b + 4ab^2)$

Solution: $3a{\cdot}a^2b + 3a{\cdot}4ab^2$

Answer: $3a^3b + 12a^2b^2$

For each of the products, fill in the correct answer in the space provided.

1. $3(x - 2)$ Answer _____

2. $-2(y - 4)$ Answer _____

3. $a(a + 4)$ Answer _____

4. $-b(b - 7)$ Answer _____

5. $-3x(x - 6)$ Answer _____

6. $4y(-y + 7)$ Answer _____

7. $-5z(2z^2 - 7z)$ Answer _____

8. $8x(-7x^2 - 8x)$ Answer _____

9. $5s^2(-2s + 7)$ Answer _____

10. $-4ab^2(-2a + 7)$ Answer _____

11. $7xy^2(3x^2y - 1)$ Answer _____

12. $-4ab^2(3a^2b + 3ab^2)$ Answer _____

13. $7a^2b(-4ab^2 + 3a^2b)$ Answer _____

14. $-2xy^3(3x^3y^3 - 2x^2y^2)$ Answer _____

15. $7x^2y(-3x^2 + 9y)$ Answer _____

16. $-3xy(-4x^2y - 5y^2x)$ Answer _____

17. $-4x(x - 7)$ Answer _____

18. $3x^2(6x - 7)$ Answer _____

30. Multiplication of a Binomial by a Binomial

Rules for Multiplication of a Binomial by a Binomial:

HINT: Multiply each term in the first binomial with each term in the second binomial.

Example 1: $(3x - 5)$ $(2x - 1)$

Solution:

(1) Multiply first term of each binomial to find the first term of the answer.

     $(3x)$ $(2x) = 6x^2$

(2) The middle term of the answer is found by adding the products of the two outside terms with the two inside terms in each binomial.

$$3x(-1) = -3x \text{ outside}$$
$$-5(2x) = \underline{-10x} \text{ inside}$$
$$-13x$$

69

(3) The last term of the answer is found by multiplying the last term of each binomial.

$$(-5) \quad (-1) = 5$$

Answer: $6x^2 - 13x + 5$

For each of the following products fill in the correct answer in the space provided.

1. $(x - 1) \quad (x + 3)$ Answer _____

2. $(x - 4) \quad (x + 5)$ Answer _____

3. $(2x - 3) \quad (2x + 4)$ Answer _____

4. $(3x - 2) \quad (2x + 1)$ Answer _____

5. $(y - 2) \quad (y + 2)$ Answer _____

6. $(y - 5) \quad (y + 5)$ Answer _____

7. $(x - 7) \quad (x + 3)$ Answer _____

8. $(x - 5) \quad (x + 6)$ Answer _____

9. $(3x - 6) \quad (3x + 6)$ Answer _____

10. $(2x - 4) \quad (3x + 8)$ Answer _____

11. $(x - 2)^2$ Answer _____

12. $(x + 5)^2$ Answer _____

31. Multiplication of Polynomials

Rules for Multiplication of Polynomials:

HINT: To find the product of two polynomials use the column method just like in long multiplication with whole numbers.

Example: $(2x + 4) (3x^2 + 2x + 2)$

Solution: $3x^2 + 2x + 2$

$$\underline{\qquad 2x + 4}$$

$$12x^2 + 8x + 8 \quad \leftarrow 4(3x^2 + 2x + 2)$$
$$\underline{6x^3 + \quad 4x^2 + 4x \qquad \leftarrow 2x(3x^2 + 2x + 2)}$$

Answer: $6x^3 + 16x^2 + 12x + 8$ Add like terms

For each of the products, fill in the correct answer in the space provided.

1. $(x + 2)$ $(x^2 + 3x + 4)$ Answer _____

2. $(y - 4)$ $(y^2 + 2y + 6)$ Answer _____

3. $(2x + 1)$ $(x^2 - 2x + 4)$ Answer _____

4. $(y + 3)$ $(y^2 - 3y + 9)$ Answer _____

5. $(x - 2)$ $(x^2 + 4x + 6)$ Answer _____

6. $(y - 3)$ $(y^2 - y + 4)$ Answer _____

7. $(5y^2 + 2y + 2)$ $(y^2 - 3y + 5)$ Answer _____

8. $(2x^2 + x + 1)$ $(x^2 - 4x + 3)$ Answer _____

9. $(y - 5)$ $(y^2 + 3y + 7)$ Answer _____

10. $(3x - 6)$ $(x^2 - 4x + 7)$ Answer _____

32. Evaluating Algebraic Expressions
Rules for Evaluating Algebraic Expressions

1. Substitute the value for the unknowns in the given expression.

2. Evaluate the roots and powers.

3. Multiply or divide from left to right.

4. If necessary, add or subtract.

Example 1:   Find the value of $3a - 1$ when $a = 6$.

Solution: $3a - 1$

            $3(6) - 1$ | Substitute $a = 6$
            $18 - 1$ | Order of operations

Answer: 17

Example 2:   Find the value of $a^2 + 7a$ when $a = -2$.

Solution: $a^2 + 7a$

$(-2)^2 + 7(-2)$

$4 + (-14)$ | Substitute a $= -2$
Order of operations

-10

Answer: -10

Example 3:   Find the value of $y^2 + 3x^2$ when $x = -1$ and $y = 2$.

Solution: $y^2 + 3x^2$

$(2)^2 + 3(-1)^2$

$4 + 3(1)$ | Substitute $x = -1$ and $y = 2$
Order of operations

$4 + 3$

7

Answer: 7

Example 4:   Find the value of $x^3y + 4x$ when $x = -2$ and $y = 5$

Solution: $x^3y + 4x$

$(-2)^3(5) + 4(-2)$

$(-8)\ \ (5) + 4(-2)$ | Substitute $x = -2$ and $y = 5$
Order of operations

$-40 + (-8)$

-48

Answer: -48

Example 5:   If $a = -3bc$, find a when $b = -2$ and $c = -3$.

Solution: $a = -3bc$

$a = -3(-2)\ (-3)$ | Substitute $b = -2$ and $c = -3$
Order of operations

$a = 6(-3)$

$a = -18$

Answer: -18

Fill in the correct answer in the space provided.

1. Find the value of $3x - 4$ when $x = -2$ Answer _____

2. Find the value of $a^2 - 7a$ when $a = 4$ Answer _____

3. Evaluate $-b^2$ when $b = -5$ Answer _____

4. Evaluate $t^2 - 7t + 6$ when $t = 8$ Answer _____

5. Find the value of $3a + 5b$ when $a = -6$ and $b = -7$ Answer _____

6. Find the value of $4c - 5d$ when $c = 2$ and $d = 3$ Answer _____

7. Find the value of $6x^2 + 7y^2$ when $x = 3$ and $y = 4$ Answer _____

8. Find the value of $x^2 + 2xy$ when $x = 3$ and $y = 4$ Answer _____

9. If $a = 3bc$, find "a" when $b = 4$ and $c = -5$ Answer _____

10. If $x = 3y^2z$, find "x" when $y = 2$ and $z = 4$ Answer _____

11. Evaluate $-3x^2$, when $x = -4$ Answer _____

12. Evaluate $7x^2 - 4y^2$, when $x = 2$ and $y = 3$ Answer _____

13. Find the value of $x^2 + 6y$, when $x = 4$ and $y = 5$ Answer _____

14. Evaluate $-4x^2$ when $x = -5$ Answer _____

33. Order of Operations in Algebra

To Simplify
HINT: First perform multiplication and then addition and subtraction.

Example 1: Simplify $3x + x(x + 6)$

Solution: $3x + x(x + 6)$
$3x + x^2 + 6x$
$x^2 + 9x$

Distribution Law of Multiplication
Addition of like terms

Answer: $x^2 + 9x$

Example 2: Simplify $3ab^2 + 4ab(6b - 7)$

Solution: $3ab^2 + 4ab(6b - 7)$
$3ab^2 + 24ab^2 - 28ab$
$27ab^2 - 28ab$

Distribution Law of Multiplication
Addition of like terms

Answer: $27ab^2 - 28ab$

Example 3:   Simplify $-3x + 2x\,(x - 7)$

Solution: $-3x + 2x\,(x - 7)$
               $-3x + 2x^2 - 14x$
               $2x^2 - 17x$

Answer: $2x^2 - 17x$

| Distribution Law of Multiplication |
| Addition of like terms |

Simplify each of the following examples and write the correct answer in the space provided.

1. $3y + y\,(y + 7)$ Answer _____

2. $2x + x\,(x + 4)$ Answer _____

3. $3x + x^2\,(x - 7)$ Answer _____

4. $4y + 4y^2\,(-y + 9)$ Answer _____

5. $6x^2 - 7x\,(x - 4)$ Answer _____

6. $-3y^2 + 2y\,(y + 2)$ Answer _____

7. $7x^2 + 2x^2\,(-x - 4)$ Answer _____

8. $3y^2 + 4y^2\,(-y - 7)$ Answer _____

9. $3a^2b - 2a\,(ab - 7)$ Answer _____

10. $x^2y - 7x\,(x^2 + 3xy)$ Answer _____

11. $3x^2y^2 - 7x\,(xy - 7xy^2)$ Answer _____

12. $-6xy - 3xy\,(-2x - 5y)$ Answer _____

13. $6a^2b - 3ab\,(7a - 2b)$ Answer _____

14. $4x^2y - 3xy(-2x - 5y)$ Answer _____

15. $5y + y\,(y + 9)$ Answer _____

16. $5x + x\,(x - 9)$ Answer _____

17. $7x^2 + 4x^2\,(-x - 7)$ Answer _____

18. $3y^2 + 5y^2\,(y + 5)$ Answer _____

19. $3x^2 - 2x\,(x + 4)$ Answer _____

20. $4y^2 + 6y^2(y + 6)$ Answer _____

74

34. Raising a Number to a Power − Exponent Rule

Exponent Rule: Raising a Number to a Power

To raise a number to a power, multiply the exponents.

$$(X^m)^n = x^{mn}$$

Example 1: $(a^4)^2$
Solution: $(a^4)^2 = a^4 \cdot a^4 = a^8$

Answer: a^8

Example 2: $(x^2y^4)^3$

Solution: $(x^2y^4)^3 = x^{2 \cdot 3}y^{4 \cdot 3} = x^6y^{12}$

Answer: x^6y^{12}

Example 3: $(5x^2y^4)^3$

Solution: $(5x^2y^4)^3 = 5^3x^{2 \cdot 3}y^{4 \cdot 3} = 125x^6y^{12}$

Answer: $125x^6y^{12}$

Fill in the letter of the correct answer in the space provided.

1. $(x^4)^3$
   (a) x (b) x^7 (c) x^{12}
   (d) x^5 (e) x^{64} Answer _____

2. $(y^{12})^2$
   (a) y^{10} (b) y^{24} (c) y^{14}
   (d) y^{144} (e) y^{36} Answer _____

3. $(x^7)^2$
   (a) x^{14} (b) x^{49} (c) x^9
   (d) x^5 (e) x^{17} Answer _____

4. $(y^{15})^4$
    (a) y^{19} (b) y^{11} (c) y^{10}
    (d) y^{24} (e) y^{60}

    Answer _____

5. $(ab^2)^3$
    (a) a^2b (b) a^3b^8 (c) ab^6
    (d) a^2b^2 (e) a^3b^6

    Answer _____

6. $(c^2d^4)^5$
    (a) $c^{10}d^{20}$ (b) $c^{15}d^7$ (c) c^3d
    (d) c^3d^9 (e) c^4d^7

    Answer _____

7. $(c^5d^7)^3$
    (a) c^2d^4 (b) $c^{15}d^7$ (c) c^5d^4
    (d) c^2d^{21} (e) $c^{15}d^{21}$

    Answer _____

8. $(4b^2)^3$
    (a) $16b^4$ (b) $64b^6$ (c) $12b^5$
    (d) $4b^5$ (e) $12b^6$

    Answer _____

9. $(2x^2y^3)^4$
    (a) $16x^8y^{12}$ (b) $16x^6y^7$ (c) $64x^8y^{12}$
    (d) $64x^2y^{12}$ (e) $16x^8y^7$

    Answer _____

35. Division of Algebraic Expressions

Exponent Rule for Division

When dividing numbers with the same base, keep the base and subtract the exponents (exponent in the numerator minus exponent in the denominator).

Example 1: $\dfrac{c^7}{c^2}$

Solution: $\dfrac{c^7}{c^2} = c^{7-2} = c^5$

Answer: c^5

Example 2: $\dfrac{a^7}{a}$

Solution: Notice $a = a^1$

$$\dfrac{a^7}{a} = \dfrac{a^7}{a^1} = a^{7-1} = a^6 \quad \text{Notice } a = a^1$$

Answer: a^6

Fill in the correct answer in the space provided.

1. $\dfrac{x^4}{x^2}$ Answer _____   5. $\dfrac{x^3}{x^7}$ Answer _____

2. $\dfrac{a^7}{a^2}$ Answer _____   6. $\dfrac{a^6}{a}$ Answer _____

3. $\dfrac{b^7}{b}$ Answer _____   7. $\dfrac{b^6}{b^2}$ Answer _____

4. $\dfrac{x^7}{y}$ Answer _____   8. $\dfrac{x^8}{y^2}$ Answer _____

Rules for Division in Algebra by a Monomial

1. Divide each part of the numerator by the denominator.

2. Determine the sign of the quotient.

3. Divide the coefficients.

4. Divide the variables by subtracting the exponents, if the bases are the same.

Example 1: $\dfrac{8a^2}{4a}$

Solution: Divide coefficients and then subtract exponents provided the bases are the same.

$$\dfrac{8a^2}{4a} = 2a$$

Answer: 2a

Example 2: $\dfrac{6b^4 - 12}{3}$

Solution: $\dfrac{6b^4 - 12}{3} = \dfrac{6b^4}{3} - \dfrac{12}{3} = 2b^4 - 4$

Answer: $2b^4 - 4$

Example 3: $\dfrac{15x^7 - 5x}{5x}$

Solution: $\dfrac{15x^7 - 5x}{5x} = \dfrac{15x^7}{5x} - \dfrac{5x}{5x} = 3x^6 - 1$

Answer: $3x^6 - 1$

Fill in the correct answer in the space provided.

1. $\dfrac{6x^2}{2x}$ Answer _____

2. $\dfrac{-4x^2}{-x^2}$ Answer _____

3. $\dfrac{6a - 12}{-3}$ Answer _____

4. $\dfrac{10y - 20}{-5}$ Answer _____

5. $\dfrac{6a^2 - 3a}{-3a}$ Answer _____

6. $\dfrac{8y^3 - 6y^2}{-4y}$ Answer _____

36. Common Factors

Factor: When two or more numbers are multiplied, each is called a factor. The process is called factoring.

Rules for Common Factors

1. Find the largest number that divides evenly into the coefficients.

2. To factor out a variable as a common factor, the variable must appear in each part of the expression. Factor out the variables to the lowest exponent that appears.

3. Divide each segment of the expression by the common factor.

Example 1: Factor completely $6x^2 - 2$

Solution: $6x^2 - 2$

STEP 1: Largest number that divides evenly into the coefficients is 2.

STEP 2: Divide each segment of the binomial by 2.

$$\frac{6x^2}{2} - \frac{2}{2} = 2(3x^2 - 1)$$

Answer: $2(3x^2 - 1)$

Example 2: Factor completely $10a^3 + 15a^2$

Solution: $10a^3 + 15a^2$

STEP 1: Largest number that divides evenly into the coefficients is 5.

STEP 2: Provided the same variable is present in each segment of the binomial, factor it out to the lowest exponent that is present (a^2).

STEP 3: Common factor is $5a^2$.

$$\frac{10a^3}{5a^2} + \frac{15a^2}{5a^2} = 5a^2(2a + 3)$$

Answer: $5a^2(2a+3)$

For each of the following examples write the common factor in the space provided.

1. $3x^2 + 6$ Answer _____

2. $12a - 24$ Answer _____

3. $6x^2 - 16$ Answer _____

4. $x^2 + 6x$ Answer _____

5. $a^4 + 7a^3$ Answer _____

6. $6x^2 + 12x$ Answer _____

7. $3x^2 - 6x$ Answer _____

8. $9x^2 - 18$ Answer _____

9. $3x^2y + 6x$ Answer _____

10. $10x^2y + 20y^2$ Answer _____

11. $18x^7y - 9x^6y^6$ Answer _____

12. $6x^3y^7 - 12x^2y^6$ Answer _____

13. $x^2 - 4x$ Answer _____

14. $6a^2 + 3b^2$ Answer _____

15. $6ab - 12a^2b^2$ Answer _____

16. $7x^2y^4 - 14x^{14}y^3$ Answer _____

17. $9x^2 + 18y^2$ Answer _____

18. $7x^2 + 14xy$ Answer _____

37. Trinomial Factors

Rules for Factoring Trinomials in Which the Coefficient of x^2 is 1

1. Factor out the highest common factor.

2. To factor one must find two numbers whose product is the last term (constant term) and the sum of the same two numbers is the middle term.

    Example 1:   Factor completely $x^2 + 7x + 10$

    Solution: Two numbers are needed whose product is 10 and the sum of the same two numbers is 7. The numbers are 5 and 2.

      Answer: $(x + 5)(x + 2)$

    Example 2:   Factor completely $x^2 - 5x - 6$

    Solution: Two numbers are needed whose product is 6 and the sum of the same two numbers with opposite signs is -5. The numbers are -6 and 1.

      Answer: $(x - 6)(x + 1)$

For each of the following example write the factors in the space provided.

1. $x^2 + 5x + 4$ Answer _____

2. $x^2 + 7x + 12$ Answer _____

3. $x^2 - 3x + 2$ Answer _____

4. $x^2 - 4x - 5$ Answer _____

5. $y^2 + y - 12$ Answer _____

6. $y^2 - 9y + 20$ Answer _____

7. $x^2 + 7x + 6$ Answer _____

8. $x^2 + 13x + 42$ Answer _____

9. $y^2 + 3y - 18$ Answer _____

10. $x^2 + 15x + 56$ Answer _____

HINT: COMMON FACTOR FIRST

11. $2x^2+6x+4$ Answer _____

12. $2y^3+4y^2+30y$ Answer _____

13. $2y^2+18y+28$ Answer _____

14. $5x^3+45x^2+100x$ Answer _____

38. Factors: The Difference of Two Squares
Rules for Factoring the Difference of Two Squares

1. The first term of the expression is a perfect square and the last term of the expression is a perfect square separated by a negative sign.

2. Factor out the highest common factor.

3. Find the square root of the first term and then find the square root of the second term. One set of factors is separated by a positive sign and the second set of factors by a negative sign.

Example 1:   Factor completely a^2-36

Solution: $a^2 - 36$

STEP 1: Determine if the expression is the difference of perfect squares.

STEP 2: Factor out common factors first.

STEP 3: Take the square root of the terms. One factor has a positive sign and the other a negative sign.

$(a+6)(a-6)$

Answer: $(a+6)(a-6)$

Example 2: Factor completely $x^2 + 36$

Solution: Not factorable—as there are no common factors and it is not a difference of perfect squares (+ sign separates the x^2 and the 36).

Answer: Cannot be factored.

Example 3: Factor completely $16a^2 - 64$

Solution: $16a^2 - 64$

$$\frac{16a^2}{16} - \frac{64}{16} = 16(a^2 - 4) \qquad \text{Common Factor.}$$

$$16a^2 - 4 = 16(a + 2)(a - 2) \qquad \text{Use method of difference of perfect squares.}$$

Answer: $16(a + 2)(a - 2)$

For each of the following examples, write the factors in the space provided.

1. $x^2 - 25$ Answer _____

2. $y^2 - 64$ Answer _____

3. $4x^2 + 36$ Answer _____

4. $x^2 - 49$ Answer _____

5. $9x^2 - 16y^2$ Answer _____

6. $49y^2 - 25x^2$ Answer _____

7. $3x^2 - 48$ Answer _____

8. $2x^2-18$ Answer _____

9. $9-x^2$ Answer _____

10. $16-y^2$ Answer _____

11. x^2-16 Answer _____

12. x^6-36 Answer _____

13. x^8-81 Answer _____

14. $9x^4-81$ Answer _____

39. Solving Linear Equations

Rules:

1. An equation is a statement of equality. There are three parts to an equation; the right side, the left side and the equal sign.

2. To solve an equation, use the Addition Property. Additive Inverse: for every number A, there is exactly one number $-A$ such that: $A + (-A) = 0$
$$X - 2 = 3$$
$$X = 5 \text{ (Add 2 to both sides).}$$

3. If necessary, in the last step, use the Multiplication Property. Multiplicative Inverse: It is the product of two numerals or two variables that $= 1$.

$$n \cdot \tfrac{1}{n} = 1$$

$$2x = 6$$

$$x = 3$$

(Multiply both sides by the multiplicative inverse of 2 which is $\tfrac{1}{2}$).

Example 1: $3x-6=12$

Solution: $3x-6=12$

             $3x-6+6=12+6$ Additive Inverse

             $\tfrac{1}{3} \cdot 3x = 18 \cdot \tfrac{1}{3}$ Multiplicative Inverse

Answer: $x=6$

Example 2: $3(x+7)=21$

Solution: $3(x+7)=21$
             $3x+21=21$ Distributive Law of Multiplication

             $3x+21-21=21+21$ Additive Inverse

             $\frac{1}{3} \cdot 3x = 42 \cdot \frac{1}{3}$ Multiplicative Inverse

Answer: $x=14$

Example 3: $3x+6+2x=31$
             $5x+6=31$ Addition of Like Terms

             $5x+6-6=31-6$ Additive Inverse

             $\frac{1}{5} \cdot 5x = 25 \cdot \frac{1}{5}$ Multiplicative Inverse

Answer: $x=5$

Example 4: $6x+1=2x+45$

             $6x+1-1=2x+45-1$ Additive Inverse

             $6x=2x+44$ Additive Inverse

             $6x-2x=2x-2x+44$

             $\frac{1}{4} \cdot 4x = 44 \cdot \frac{1}{4}$ Multiplicative Inverse

             $x=11$

Answer: $x=11$

Solve each of the following equations: Write the correct answer in the space provided.

1. $x-5=2$ Answer _____

2. $x+7=4$ Answer _____

3. $2x=6$ Answer _____

4. $-3y=9$ Answer _____

5. $2y-7=9$ Answer _____

6. $-3x-6=9$ Answer _____

7. $-3x+6=12$ Answer _____

8. $4-x=0$ Answer _____

9. $3y+4=2y+6$ Answer _____

10. $y-7=1$ Answer _____

11. $2x-6=-12$ Answer _____

40. Solving Equations with Fractions

RULES:

1. Multiply each and every number and variable on both sides of the equation by the lowest common denominator (LCD).

2. Now follow the rules used for solving equations.

Example 1: $\frac{a}{6} + 12 = \frac{a}{3}$

Solution: The LCD for $\frac{a}{3}$ and $\frac{a}{6}$ is 6.

Multiply both sides by 6.

$\frac{a}{6} + 12 = \frac{a}{3}$

$6 \cdot \left(\frac{a}{6}\right) + 6(12) = 6\left(\frac{a}{3}\right)$ Use the Distributive Law of Multiplication.

$a+72=2a$

$a-a+72=2a-a$ Additive Inverse.

$72=a$

Answer: $a=72$

Example 2: $\frac{x}{3} + \frac{x}{4} = 12$

Solution: The LCD for $\frac{x}{3}$ and $\frac{x}{4}$ is 12.

85

Multiply both sides by 12.

$\frac{x}{3} + \frac{x}{4} = 12$

$12 \cdot \left(\frac{x}{3} + \frac{x}{4}\right) = 12 \cdot 12$ Use the Distributive
Law of Multiplication.

$4x + 3x = 144$ Addition of like terms.

$\frac{1}{7} \cdot 7x = 144 \cdot \frac{1}{7}$ Multiplicative Inverse

$x = \frac{144}{7}$

Answer: $x = \frac{144}{7}$

Solve each of the following equations. Write the correct answer in the space provided.

1. $\frac{x + 1}{2} = \frac{3x + 6}{4}$ Answer _____

2. $\frac{x - 4}{3} = \frac{2x + 1}{5}$ Answer _____

3. $\frac{x}{2} + \frac{x}{4} = 3$ Answer _____

4. $\frac{x}{3} + \frac{x}{9} = 4$ Answer _____

5. $\frac{x}{12} + 1 = \frac{x}{10}$ Answer _____

6. $\frac{x}{2} + 3 = \frac{x}{4}$ Answer _____

7. $\frac{u}{10} = \frac{u}{5} - 2$ Answer _____

8. $\frac{x}{7} + 2 = \frac{x}{14}$ Answer _____

9. $\frac{x}{2} + \frac{x}{3} = 5$ Answer _____

41. Solving Literal Equations

RULES:

1. A literal equation is an equation with two or more variables.

2. Isolate the variable to be solved.

3. Follow the rules used to solve linear equations.

Example 1:   If $3y - 7 = x$, solve for y.

Solution: To solve for y, the variable y must be isolated.

$$3y - 7 = x$$

$$3y - 7 + 7 = x + 7 \qquad \text{Additive Inverse}$$

$$\tfrac{1}{3} \cdot 3y = \tfrac{1}{3} \cdot (x + 7) \qquad \text{Multiplicative Inverse}$$

$$y = \frac{x + 7}{3}$$

Answer: $y = \frac{x + 7}{3}$

Example 2:   If $bx + a = 3c$, solve for x.

Solution: Isolate x on the left side. Use Additive Inverse by adding $-a$ to both sides of the equation.

$$bx + a = 3c$$

$$bx + a - a = 3c - a \qquad \text{Additive Inverse}$$

$$\tfrac{1}{b} \cdot bx = \tfrac{1}{b} \cdot (3c - a) \qquad \text{Multiplicative Inverse}$$

$$x = \frac{3c - a}{b}$$

Answer: $x = \frac{3c - a}{b}$

Solve each of the following equations for the variable indicated:
Write the correct answer in the space provided.

1. $x - y = 7$ Solve for x: Answer _____

2. $x - y = 3$ Solve for y: Answer _____

3. $a - 7b = 0$ Solve for a: Answer _____

4. $a - 2b = 0$ Solve for b: Answer _____

5. $3x - 2 = y$ Solve for x: Answer _____

6. $4x - 5y = 6$ Solve for y: Answer _____

7. $\frac{x}{2} + 4 = y$ Solve for x: Answer _____

8. $\frac{x}{3} - 7 = y$ Solve for x: Answer _____

9. $x + y + z = 6$ Solve for z: Answer _____

10. $3x - 4y + 7z = 9$ Solve for y: Answer _____

11. $\frac{a - b}{4} = c$ Solve for a: Answer _____

12. $3(x + y) = 6$ Solve for x: Answer _____

13. $-2(x - y + z) = 12$ Solve for x: Answer _____

42. Solving Verbal Problems − Number Problems

RULES

English	Algebra
The sum of x and y	x + y
The product of x and y	xy
The quotient of x and y is	$\frac{x}{y}$
A number	x
6 more than a number	6 + x
4 times a number	4x
The difference of x and y	x − y

Example: The sum of twice a number and four is eight. Find the number.

Solution:

1. Let x = the number.

2. The sum of twice a number and four is represented as: 2x + 4.

3. The word "is" translates as =.

4. Eight is 8.

5. Equation: 2x + 4 = 8

6. Solve for x.

2x + 4 = 8	Add −4 to both sides Additive Inverse
2x = 4	Multiply both sides by $\frac{1}{2}$ Multiplicative Inverse
x = 2	

Answer: x = 2

Solve the following problems. Write the correct answer
in the space provided.

1. The sum of twice a number and six is twelve.
   Find the number.   Answer _____

2. Four times the sum of a number and seven is thirty.
   Find the number.   Answer _____

3. One number is four less than another. Their sum is
   eight. Find both numbers.   Answer _____

4. The sum of twice a number and eight is four.
   Find the number.   Answer _____

5. If twice the difference of a number and three were
   decreased by five, the result would be three.
   Find the number.   Answer _____

6. One number is 6 less than another. Their sum is
   twelve. Find the larger number.   Answer _____

7. Five times the sum of a number and six is forty.
   Find the number.   Answer _____

8. The sum of twice a number and nine is fifteen.
   Find the number.   Answer _____

43. Ratios and Proportions − Verbal Problems

1. Ratio: If x and y are any two number where
y ≠ 0, then the ratio of x and y is:

$$\frac{x}{y}$$

2. Proportion: Is a statement that two ratios are equal.

Means − Extremes Property

If a, b, c and d are real numbers, when b ≠ 0 and d ≠ 0 then:

if $\frac{a}{b} = \frac{c}{d}$

then ad = bc

In words: In any proportion, the product of the means is equal
to the product of the extremes.

Example 1:   Solve $\frac{3}{x} = \frac{6}{7}$

Solution:

1. $\frac{3}{x} = \frac{6}{7}$ Extremes are 3 and 6.
   Means are x and 7.

2. 6x = 21   Products of means = product
   of extremes.

3. x = $\frac{21}{6}$ Multiply both sides by $\frac{1}{6}$.
   Multiplicative Inverse.

4. x = $\frac{7}{2}$ Reduce.

Answer: x = $\frac{7}{2}$

Example 2:   A baseball player gets 6 hits in the first 18 games. If he continues hitting at the same rate, how many hits will he get in the first 45 games?

Solution:

1. Ratio 1 = Ratio 2

2. $\frac{\text{Hits}}{\text{Games}} = \frac{\text{Hits}}{\text{Games}}$

3. $\frac{6}{18} = \frac{x}{45}$

4. $270 = 18x$ Product of extremes = product of means.

5. $15 = x$ Multiply both sides by $\frac{1}{18}$ Multiplicative Inverse.

Answer: $x = 15$

Solve each of the following proportions. Write the correct answer in the space provided.

1. $\frac{x}{3} = \frac{6}{9}$ Answer _____

2. $\frac{4}{x} = \frac{2}{3}$ Answer _____

3. $\frac{15}{60} = \frac{60}{x}$ Answer _____

4 $\frac{2}{7} = \frac{x}{14}$ Answer _____

5 $\frac{x + 2}{3} = \frac{2}{5}$ Answer _____

6. $\frac{x + 1}{4} = \frac{2}{9}$ Answer _____

Solve each of the following verbal problems by the method of ratios and proportions. Write the correct answer in the space provided.

1. If 200 grams of ice cream contain 26 grams of fat, how much fat is in 700 grams of ice cream?

   Answer _____

2. A map is drawn so that 2 inches represents 350 miles. If the distance between 2 cities is 875 miles, how far apart are they on the map?

   Answer _____

3. An airplane flies 1,950 miles in 6 hours. How far will it travel in 7 hours?

   Answer _____

4. A man drives his car 630 miles in 10 hours. At this rate how far will he travel in 12 hours?

   Answer _____

5. A 5 ounce serving of grapefruit juice contain 125 grams of water. How many grams of water are there in an 8 ounce serving of grapefruit juice?

   Answer _____

6. A basketball player makes 8 out of 12 free throws in the first game of the season. If she shoots with the same accuracy in the second game, how many of the 15 free throws she attempts will she make?

   Answer _____

44. Solving Simultaneous Equations

Simultaneous Equations are two Equations with two different variables.

RULES: CATEGORY #1

1. **If the equation is in the following form:**

   x + y = 6
   x − y = 2

   Add both equations to eliminate one of the variables. Then solve for the other variable.

2. **Substitute the value found for the variable in either equation and solve the equation for the other variable.**

Example 1: Solve the system of equations $x + y = 6$
and $x - y = 2$.

Solution:

$x + y = 6$
$\underline{x - y = 2}$
$\quad 2x = 8$ — Eliminate the $+y$ and the $-y$ in both equations by addition.

$\frac{1}{2} \cdot 2x = \frac{1}{2} \cdot 8$ — Solve for x by using the multiplicative inverse of 2 which is $\frac{1}{2}$.

$x = 4$

$4 + y = 6$ — Substitute 4 for x in either equation to find y.

$4 - 4 + y = 6 - 4$ — Additive Inverse of $+4$ is -4.

$Y = 2$

Answer: $x = 4, y = 2$

Solve each of the following simultaneous equations. Fill in
the correct answer in the space provided.

1. $x - y = 2$
    $\underline{x + y = 4}$ Answer _____

2. $a - b = 5$
    $\underline{a + b = 3}$ Answer _____

3. $x - 2b = 5$
    $\underline{x + 2b = 3}$ Answer _____

4. $-2x + y = 6$
    $\underline{\ 2x + 2y = 6}$ Answer _____

5. $a - b = 6$
    $\underline{a + b = 2}$ Answer _____

RULES: CATEGORY#2

1. **If the equation is in the following form:**

    $2x - y = 6$
    $\underline{x + 3y = 3}$

    Solve by the following procedure:

2. **Multiply each and every number on both sides of
    the first equation by 3 or multiply the second equation
    by −2.**

3. **Add both equations to eliminate one set of variables and
    solve for the other variable.**

4. **Substitute the value found for the variable in one of the
    equations and solve for the other variable.**

Example 1:   Solve the system of equations $2x - y = 6$ and $x + 3y = 3$.

Solution: $2x - y = 6$
$\underline{x + 3y = 3}$
$6x - 3y = 18$ Multiply top equation
on both sides by 3.

$\underline{x + 3y = 3}$ Eliminate the $-3y$ and the $+3y$
$7x = 21$ in both equations by addition.

$\frac{1}{7} \cdot 7x = \frac{1}{7} \cdot 21$ Multiplicative inverse of 7 is $\frac{1}{7}$.

$x = 3$

$3 + 3y = 3$ Substitute 3 for x in either
equation to find y.

$3 - 3 + 3y = 3 - 3$ Additive Inverse of $+3$ is -3.

$3y = 0$

$\frac{1}{3} \cdot 3y = \frac{1}{3} \cdot 0$ Multiplicative Inverse of 3 is 1/3.

Answer: $x = 3, y = 0$

Solve each of the following simultaneous equations. Fill in the correct answer in the space provided.

1. $2x - y = 8$
$\underline{x + 3y = 4}$ Answer _____

2. $2x - 6y = 4$
$\underline{x + 2y = 3}$ Answer _____

3. $4x - y = 6$
$\underline{x + 2y = 15}$ Answer _____

4. $2a - 3b = -3$
$\underline{a + 4b = 4}$ Answer _____

5. $3x - 2y = -8$
$\underline{x + 4y = 2}$ Answer _____

96

1. **Some simultaneous equations can be solved by substitution.**

   $2x + 3y = 7$
   $\underline{\qquad\; y = x - 1}$

   In the above problem, substitute "x − 1" for "y" in the top equation and then solve for x.

   $2x + 3(x - 1) = 7$

2. **Substitute the value found for the variable in either one of the equations, and solve for the other variable.**

Example 1:   Solve the system of equations $2x + 3y = 7$
              and $y = x - 1$.

Solution: $2x + 3y = 7$
              $\underline{\qquad\; y = x - 1}$

$2x + 3(x - 1) = 7$	Substitute x − 1 for y in the top equation.
$2x + 3x - 3 = 7$	Distributive Law of Multiplication.
$5x - 3 = 7$	Addition of like terms.
$5x - 3 + 3 = 7 + 3$	Additive Inverse of −3 is + 3.
$5x = 10$	
$\frac{1}{5} \cdot 5x = \frac{1}{5} \cdot 10$	Multiplicative inverse of 5 is $\frac{1}{5}$.
$x = 2$	
$y = 2 - 1$	Substitute 2 for x in the bottom equation to find y.
$y = 1$	

Answer: $x = 2, y = 1$

Solve each of the follow simultaneous equations. Fill in the correct answer in the space provided.

1. $3x + 4y = 6$
    $\underline{\hspace{1cm} y = \ x - 1}$ Answer _____

2. $4x + 2y = 6$
    $\underline{\hspace{1.5cm} x = y}$ Answer _____

3. $2x + 3y = 7$
    $\underline{\hspace{1cm} y = x + 4}$ Answer _____

4. $2x + 4y = 6$
    $\underline{\hspace{1cm} x = y + 3}$ Answer _____

5. $3x + 6y = 6$
    $\underline{\hspace{1cm} x = 2y + 6}$ Answer _____

6. $4x + 6y = 16$
    $\underline{\hspace{1cm} y = 2x + 8}$ Answer _____

7. $2x + y = 6$
    $\underline{\hspace{1.5cm} y = x}$ Answer _____

45. Solving Quadratic Equations

RULES

1. Any equation that can be put into the form $ax^2 + bx + c = 0$ is called a quadratic equation. There are two solutions for each equation.

2. Set equation equal to zero.

3. Factor.

4. Set each factor equal to zero.

5. Solve for each of the factors.

Example 1: $x^2 + 7x = -6$

Solution:

1. Set equation equal to zero.
   $x^2 + 7x + 6 = 0$

2. Factor:
   $(x + 6)\ \ (x + 1) = 0$

3. Set each factor equal to zero.
   $x + 6 = 0$ or $x + 1 = 0$

4. Solve:
   $x = -6$ or $x = -1$

Answer: $(-6, -1)$ solution set

Solve each of the following quadratic equations. Write the solution set in the space provided.

1. $y^2 + 7y + 10 = 0$ Answer _____

2. $x^2 + 12x + 11 = 0$ Answer _____

3. $x^2 + 5x = -4$ Answer _____

4. $y^2 - 9y = -20$ Answer _____

5. $x^2 + 2x = 3$ Answer _____

6. $x^2 = 16$ Answer _____

7. $x^2 - x = 6$ Answer _____

8. $x^2 = 25$ Answer _____

9. $x^2 - 7x = -12$ Answer _____

10. $y^2 + 4y = -3$ Answer _____

11. $x^2 + 5x = -6$ Answer _____

12. $x^2 + 8x = -15$ Answer _____

46. Graph of a Straight Line

A graph is a picture used to show some information. The center of the graph is the origin. At the origin, the center of the graph, both the x and the y values are zero. The horizontal line represents the x axis and the vertical line represents the y axis.

To draw a graph of a straight line, one must find points that are solutions to the equation. For example, in the equation x = y+1, one must assume values of one of the variables and solve for the other variable. If one assumes y = 1, substitute the value in the equation and solve for x; (x = 2). Choose several values for y and find the values for x. Locate them on the graph. The points should connect to a straight line.

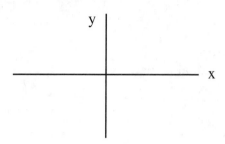

RULES TO GRAPH

1. **Choose at Least Three Different Values for One Variable.**

2. **Substitute in the Equation and Find the Value of the Other Variable.**

3. **Plot the (x, y) Values for Each Point.**

4. **Connect the Points as a Straight Line.**

In writing a point, use the following notation: (x, y). The x value is written first, and then the y value separated by a comma and enclosed by a parenthesis. The point $(-3, -5)$ has an x value of -3 and a y value of -5

47. To Determine if a Point Lies on a Graph

RULES: To Determine if a Point Lies on a Graph

1. **Substitute in the equation the values given.**

2. **If both sides of the equation have the same value, the point is a solution to the equation.**

Example 1:   Which of these points lies on the graph $4x + y = 6$?

        (a) $(2,7)$ (b) $(-3, -5)$
        (c) $(1,2)$ (d) $(6,7)$

Solution: Solve equation by substituting the value of each point for x and y in the equation. Start with choice (A) until you find the correct answer.

1.   Substitute $(2, 7)$ 2 for x and 7 for y.

$$4x + y = 6$$
$$4(2) + 7 = 6$$
$$8 + 7 = 6 \qquad \textbf{False}$$

2.   Substitute $(-3, -5)$ -3 for x and -5 for y.

$$4x + y = 6$$
$$4(-3) + (-5) = 6$$
$$-12 - 5 = 6 \qquad \textbf{False}$$

3.   Substitute $(1, 2)$ 1 for x and 2 for y.

$$4x + y = 6$$
$$4(1) + 2 = 6$$
$$4 + 2 = 6 \qquad \textbf{True}$$

Answer: $(1,2)$

Fill in the letter of the correct answer in the space provided.

1.   Which one of the following points lies on the graph:
$y = 3x + 1$?
  (a) $(3, 4)$ (b) $(-2, -7)$
  (c) $(-3, -4)$ (d) $(2, 7)$ Answer _____

2.   Which one of the following points lies on the graph:
$2x + 3y = 5$?
  (a) $(3, 4)$ (b) $(1, 1)$
  (c) $(-3, -4)$ (d) $(2, 9)$ Answer _____

3. Which one of the following points lies on the graph:
    y = x − 2?
    (a) (3, 7)   (b) (2, 1)
    (c) (4, 2)   (d) (7, 3)
    (e) (7, 4)   Answer _____

4. Which one of the following points lies on the graph:
    6x − 5y = 1?
    (a) (2, 4)   (b) (1, 1)
    (c) (6, 2)   (d) (3, 4)   Answer _____

5. Which one of the following points lies on the graph:
    x − y = 1?
    (a) (5, 7)   (b) (4, 6)
    (c) (3, 2)   (d) (−4, −3)   Answer _____

6. Which one of the following points lies on the graph:
    y = 4 − x?
    (a) (3, 4)   (b) (4, 0)
    (c) (−2, 4)   (d) (−4, −5)   Answer _____

7. Which one of the following points lies on the graph:
    5x + y = 0?
    (a) (3, 4)   (b) (1, −5)
    (c) (−2, 4)   (d) (−4, −3)   Answer _____

48. To find "x" Intercept and the "y" Intercept

RULES: Used to Find Intercepts	
x—Intercept:	Is the point where the line crosses the x axis.
y—Intercept:	Is the point where the line crosses the y axis.

1. To find the x intercept, set the y value equal to zero.

    For example: 2x + 3y = 6

Solution: $2x + 3y = 6$

$y = 0$

Then, $2x = 6$

$x = 3$

Answer: $x = 3$

2. To find the y intercept, set the x value equal to zero.

For Example: $3x + 7y = 14$

Solution: $3x + 7y = 14$

$x = 0$

Then, $7y = 14$

$y = 2$

Answer: $y = 2$

Fill in the letter of the correct answer in the space provided.

1. Find the x intercept for the equation $2x - y = 6$. Answer _____

2. Find the x intercept for the equation $-2x - 3y = -6$. Answer _____

3. Find the x intercept for the equation $-7x - 7y = 14$. Answer _____

4. Find the x intercept for the equation $3x + 6y = 12$. Answer _____

5. Where does the graph $2x - 3y = 6$ intercept the x axis? Answer _____

6. Where does the graph $-2x - 7y = 14$ intercept the y axis? Answer _____

7. Where does the graph $3x + 6y = 1$ intercept the x axis? Answer _____

8. Where does the graph $-4x - 6y = -3$ intercept the y axis? Answer _____

49. To Write an Equation of a Straight Line Knowing the "x" Intercept and the "y" Intercept

1. Use the formula:

   $\frac{x}{a} + \frac{y}{b} = 1$

   where "a" is the value of the "x" intercept and "b" is the value of the "y" intercept.

2. Remove the fractions by using the lowest common multiplier (LCD).

Example 1:   Write an equation for the line.

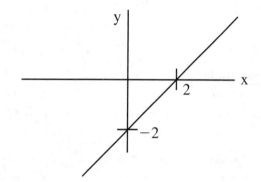

Solution: $\frac{x}{a} + \frac{y}{b} = 1$ | Write the formula. |

               $\frac{x}{2} + \frac{y}{-2} = 1$

                                                     | Substitute
                                                       a = x − intercept = 2
                                                       b = y − intercept = −2. |

               $2(\frac{x}{2}) + 2(\frac{y}{-2}) = 2(1)$

                                                     | Use LCD = 2 to clear
                                                       fractions by multiplying
                                                       both sides by 2. |

               $x - y = 2$

Answer: $x - y = 2$

Write the correct answer in the space provided.

1. Write an equation for the line:

a.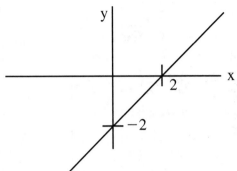

b.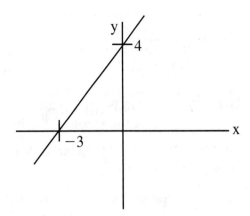

Answer _____

Answer _____

2. Write an equation for the line:

a.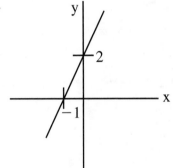

b.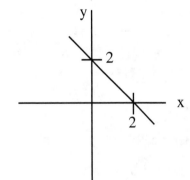

Answer _____

Answer _____

50. To Find the Slope Given Two Points

Example: Find the slope of the line between the points
(4, 3) (6, 5).

Solution:

    1. Label the points:

       $(x_1, y_1) = (4, 3)$

       $(x_2, y_2) = (6, 5)$

    2. Substitute the formula and evaluate.

$$m = \frac{y_2 - y_1}{x_2 - x_1} = \frac{5 - 3}{6 - 4} = \frac{2}{2} = 1$$

Answer: $m = 1$

Find the slope for each of the following sets of points. Write the correct answer in the space provided.

1. $(3, 1), (5, 4)$ Answer _____

2. $(2, 1), (5, 5)$ Answer _____

3. $(1, 3), (5, 2)$ Answer _____

4. $(-1, -1), (-2, -2)$ Answer _____

5. $(1, -3), (4, 2)$ Answer _____

6. $(2, -4), (3, -1)$ Answer _____

7. $(4, 5), (5, 6)$ Answer _____

8. $(-3, -2), (1, 3)$ Answer _____

9. $(2, -5), (3, -2)$ Answer _____

10. $(-3, 3), (3, -1)$ Answer _____

51. Slope Intercept Form of the Equation of a Line

The equation of a line with slope "m" and y intercept, "b" is always given by the following $y = mx + b$.

RULES

1. **Write the equation**

   Y = mx + b

2. **Substitute in the equation the value of the slope and the value of the y-intercept.**

Example: Write the equation of a line with slope 2 and y-intercept 5.

Solution: Substitute m = 2 and b = 5 in the following equation:

         $y = mx + b$
         $y = 2x + 5$

Answer: $y = 2x + 5$

For each of the following problems, write an equation: Write the correct answer in the space provided.

1. $m = 4, b = 1$ Answer _____

2. $m = 3, b = -6$ Answer _____

3. $m = -2, b = 4$ Answer _____

4. $m = -5, b = -6$ Answer _____

5. $m = -1, b = -1$ Answer _____

6. $m = 2, b = 2$ Answer _____

7. $m = \frac{2}{3}, b = 6$ Answer _____

8. $m = -\frac{2}{5}, b = \frac{3}{4}$ Answer _____

52. To Find the Slope and the "y" Intercept Given an Equation

---

RULES

1. **Write the equation in the form y = mx + b (Slope − Intercept form).**

2. **Solve the equation for y.**

Example: Find the slope and the y-intercept for $4x + 2y = 8$.

Solution: Write the equation in the form $y = mx + b$ (solve for y).

1. $4x + 2y = 8$ Original equation.

2. $2y = -4x + 8$ Add a $-4x$ to each side. Additive Inverse

3. $\frac{1}{2}(2y) = \frac{1}{2} \cdot (-4x + 8)$ Multiplicative Inverse Multiply by 1/2.

4. $y = -2x + 4$ Simplify.

5. Slope $= m = -2$
   y-intercept $= b = 4$

Answer: $m = -2, b = 4$.

Find the slope and the y−intercept for each of the following equations:
Write the correct answer in the space provided.

1. $4x + y = 6$ Answer _____

2. $3x + y = 4$ Answer _____

3. $2x − y = 5$ Answer _____

4. $5x − y = 6$ Answer _____

5. $6x + 3y = 9$ Answer _____

6. $2x + 4y = 6$ Answer _____

7. $x − y = 2$ Answer _____

8. $2x + 3y = 6$ Answer _____

53. An Introduction to Geometry

Geometry is the study of shapes, sizes and relationships of figures such as squares, rectangles, triangles and circles. Important geometric definitions include the following:

Point: A point has no length nor width. It only marks a position. It is represented by a dot. A capital letter is placed near the dot.

Line: A line is an infinite set of points. It has length but no width.

The above line can be referred to as line RS or SR.

Plane: A plane is a flat surface that extends in all directions. It has no end.

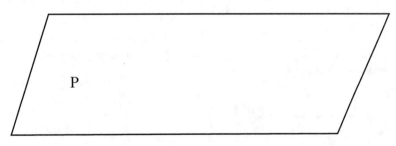

This figure represents plane P. Technically it is only part of a plane and it has no boundaries.

Information about Lines

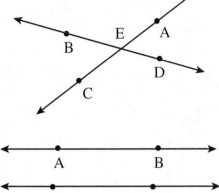

Intersecting Lines are two straight lines that meet or cross each other. They intersect at only one point.

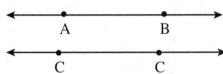

Parallel Lines are two lines in the same plane that never intersect.

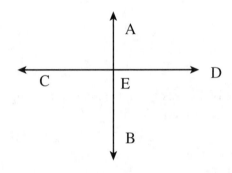

Perpendicular Lines are two straight lines that intersect to form right angles (90°). If AB is perpendicular to CD, then the four angles at E are right angles.

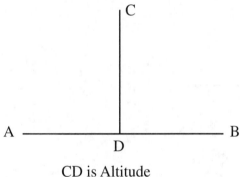

CD is Altitude

Altitude is a line drawn to another line forming right angles. An altitude of a triangle is a line drawn from a vertex perpendicular to the opposite side. It is a measure of the height of a geometric figure, such as a triangle or a rectangle.

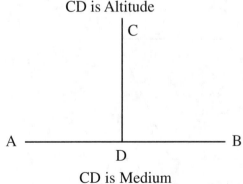

CD is Medium

Medium is a line drawn to another line that divides it into two equal line segments. In a triangle, it is a line drawn from a vertex to the midpoint of the opposite side.

54. Angle Theorems and Applications

ANGLES

Angles: An angle is the union of two rays that extend from a common end point.

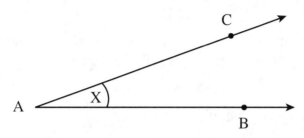

The symbol for an angle is $\angle$. The vertex is at "A". To name an angle, the vertex letter must be in the middle, if three letters are used. There are four ways to name the above angle.

1. $\angle$ CAB   2. $\angle$ BAC

3. $\angle$ A   4. $\angle$ X

The **Vertex** is the intersecting point of two sides. Angles are measured in **degrees.**

Right Angle: Right angle is an angle that measures exactly 90°.

Straight Angle: Straight angle is an angle that measures exactly 180°.

Obtuse Angle: Obtuse angle is an angle that measures less than 180° and more than 90°.

Acute Angle: Acute angle is an angle that measures less than 90°.

Bisector: Bisector is a ray that divides the angle into two equal segments.

ANGLE THEOREMS

Supplementary Angles: Supplementary angles are two or more angles that add up to 180°.

Complementary Angles: Complementary angles are two or more angles that add up to 90°.

Vertical Angles: Vertical angles are formed when two lines intersect. The opposite angles are equal. They are called vertical angles.

Sample Example #1:   An angle is eight times its supplement. Find both angles.

Solution:

Let x = smaller angle.

Let 8x = larger angle.

8x + x = 180° definition of Supplementary.

9x = 180° Add like terms.

x = 20° Answer Smaller angle.

8x = 160° Answer Larger angle.

Sample Example #2:   An angle is 60° more than its complement. Find the larger angle.

Solution:

Let x = smaller angle.

Let x + 60° = larger angle.

x + x + 60 = 90° definition of Complementary.

2x + 60° = 90° Add like terms.

2x = 30° Solve.

x = 15°

x + 60° = 75° Answer Larger angle.

Answer each of the following questions. Write the correct answer in the space provided.

1. Find the complement of an angle that is 29°.　　　　Answer _____

2. Find the supplement of an angle that is 59°.　　　　Answer _____

3. An angle is five times its supplement. Find the
    larger angle.　　　　　　　　　　　　　　　　　　Answer _____

4. An angle is four times its complement. Find
    both angles.　　　　　　　　　　　　　　　　　　Answer _____

5. An angle is 60° more than its supplement. Find
    both angles.　　　　　　　　　　　　　　　　　　Answer _____

6. An angle is 12° less than its complement. Find
    the smaller angle.　　　　　　　　　　　　　　　　Answer _____

7. An angle exceeds its supplement by 60°. Find
    the larger angle.　　　　　　　　　　　　　　　　Answer _____

8. An angle is twice its supplement. Find both angles.　　Answer:_____

55.　Triangle Theorems and Applications

TRIANGLE THEOREMS

A triangle is a polygon with three sides and three angles. (A polygon is
a closed figure formed by three or more sides) The symbol for a triangle is $\triangle$.

Equilateral Triangle:　　　is a triangle with three equal sides
　　　　　　　　　　　　　　of equal lengths.

Isosceles Triangle:　　　　is a triangle with two sides of
　　　　　　　　　　　　　　equal length.

Scalene Triangle:　　　　　is a triangle with no two sides of
　　　　　　　　　　　　　　equal length.

Right Triangle:　　　　　　is a triangle with a right angle.

The sum of the angles of a triangle is 180°

Sample Example #1:   The vertex angle of an isosceles triangle is 40°.
                        Find the number of degrees in each base angle.

Solution:

Let x = the number of degrees in each base angle.

$x + x + 40° = 180°$ the sum of the angles of a triangle.

$2x + 40° = 180°$. Add like terms.

$2x = 140°$ Solve.

$x = 70°$

Answer: Each base angle is 70°.

Sample Example #2:   The three angles of a triangle are in the ratio of
                        3:4:5. Find the number of degrees in each angle.

Solution:

Let 3x, 4x and Sx represent three angles.

$3x + 4x + 5x = 180°$ sum of the angles of a triangle.

$12x = 180°$ Add like terms.

$x = 15°$. Solve the equation.

$3x = 45°$

$4x = 60°$

$5x = 75°$

Answer: The three angles are 45°, 60°, and 75°.

Answer each of the following questions: Write the correct answer in the space provided.

1. The vertex angle of an isosceles triangle is 70°.
   Find the value of each base angle.   Answer _____

2. The three angles of a triangle are in the ratio of 1:2:3.
   Find the value of the larger angle.   Answer _____

3. The vertex angle of an isosceles triangle is 80°.
   Find the value of each base angle.   Answer _____

4. The vertex of an isosceles triangle is twice the base
   angle. Find the value of the vertex angle.   Answer _____

56. Circle Theorems – Circumference – Area

Circle Theorems:

A circle is a plane figure bounded by a curved line, every point of which is the same distance from the center of the figure. The **circumference** of a circle is the line that forms its outer boundary. It is like the perimeter of a circle. A radius of a circle is a line segment joining the center to any point on the circumference. A **diameter** is a line segment joining two points on the circumference and passing through the center. A diameter is equal to two radii.

$$C = 2\pi r \text{ circumference.}$$
$$A = \pi r^2 \text{ area}$$
$$\pi = 22/7 = 3.14$$

Sample Example #1:   Find the circumference of a circle whose diameter is 14 cm.

Solution:

The radius of a circle is half the diameter.
Two radii make up on diameter.

$C = 2\pi r$ Formula

$C = 2 (22/7)(7)$ Substitute

$C = 44$ cm. Evaluate

Answer: The circumference of the circle is 44 cm.

Sample Example #2:   Find the area of a circle whose radius is 6 cm.

Solution:

$A = \pi r^2$

$A = 3.14 (6)^2$ Substitute

$A = 3.14 (36)$ Evaluate

$A = 113$ sq. cm.

Answer: The area of the circle is 113 sq. cm.

Answer each of the following questions. Write the correct answer in the space provided.

1. Find the circumference of a circle whose radius
    is 14 cm.   Answer _____

2. Find the radius of a circle whose circumference
    is 11 cm.   Answer _____

3. Find the area of a circle whose diameter is 14 cm.   Answer _____

4. Find the circumference of a circle whose radius
    is 7 cm.   Answer _____

5. Find the radius of a circle whose circumference
    is 22 cm.   Answer _____

6. Find the area of a circle whose radius is 8 cm.   Answer _____

57. Area and Perimeter of Geometric Figures

Important Geometric Forms:

Polygon: is a closed figure formed by three or more straight lines, all of which intersect in a plane.

Triangle: is a polygon with three sides and with three angles.

Quadrilateral: is a polygon with four sides.

Parallelogram: is a quadrilateral in which both pairs of opposite sides are parallel.

Rectangle: is a parallelogram in which all angles are right angles.

Square: is a rectangle in which all sides are equal.

Rhombus: is a parallelogram in which all the sides are equal in length.

Trapezoid: is a quadrilateral in which only one pair of opposite sides are parallel.

Important Formulas: Area and Perimeter

Area: is the amount of the plane enclosed by a polygon.

Perimeter: is the distance around the polygon.

AREA OF A RECTANGLE

Area = length $\times$ width
$A = l \times w$

AREA OF A SQUARE

Area = side squared
$A = S^2$

AREA OF A TRIANGLE

Area = $\frac{1}{2}$ base $\times$ height

$A = \frac{1}{2} b \times h$

AREA OF A TRAPEZOID

Area of Trapezoid = $\frac{1}{2}$ height
(Sum of the bases)

$A = \frac{1}{2} h (b_1 + b_2)$

PERIMETER OF A RECTANGLE

Perimeter = $2 \times$ length $+ 2 \times$ width

$P = 2l + 2w$

PERIMETER OF A SQUARE

Perimeter = 4 times the side

$P = 4S$

Sample Example #1:   Find the area of a rectangle whose length
                       is 10 cm. and whose width is 5 cm.

Solution:

    $A = 1 \times w$ Formula

    $A = 10 \times 5$ Substitute

    $A = 50$ sq. cm. Evaluate

Answer: The area is 50 sq. cm.

Sample Example #2:   Find the perimeter of a square whose side
                             is 14 inches.

Solution:

    P = 4S Formula

    P = 4(14)Substitute

    P = 56 in. Evaluate

Answer: The perimeter is 56 in.

Sample Example #3:   Find the side of a square whose area
                             is 64 sq. cm.

Solution:

    $A = S^2$ Formula

    $64 = s^2$ Substitute

    $8 = s$ Evaluate (take the square root of
                     both sides)

    Answer: The side is 8 cm.

Sample Example #4:   Find the area of a trapezoid whose altitude
                             is 8 cm. and whose bases are 12 cm. and 14 cm.

Solution:

    $A = \frac{1}{2} h (b_1 + b_2)$ Formula

    $A = \frac{1}{2} 8 (12 + 14)$ Substitute

    A = 4 (26) Evaluate

    A = 104 sq. cm.

    Answer: The area is 104 sq. cm.

Answer each of the following questions. Write the correct answer in the space provided.

1. Find the perimeter of a rectangle whose dimensions
   are 14 cm. and 12 cm.   Answer _____

2. If the perimeter of a square is 48 meters, what is the
   value of the side of a square.   Answer _____

3. The area of a triangle is 48 sq. cm. If the base is 24 cm.,
   find the value of the altitude.   Answer _____

4. Find the side of square whose area is 64 sq. cm.   Answer _____

5. Find the area of square whose perimeter is 24 feet.   Answer _____

6. The area of a trapezoid is 72 sq. cm. and the sum
   of its bases is 36 cm. Find the altitude.   Answer _____

58. Pythagorean Theorem

RULES FOR THE PYTHAGOREAN THEOREM

1. This theorem applies only to a right triangle. A right triangle
   is a triangle with a 90 degree angle. The side opposite the
   right angle is called the hypotenuse. The other two sides
   are called legs. The hypotenuse is the longest side.

2. This theorem states:

   $Leg^2 + Leg^2 = Hypotenuse^2$

   $a^2 + b^2 = c^2$

Example 1:   In the right triangle shown, find x.

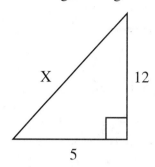

Solution:

$a^2 + b^2 = c^2$

$12^2 + 5^2 = x^2$

| Write the formula. |
| Substitute. |

$144 + 25 = x^2$

$\sqrt{169} = \sqrt{x^2}$

| Perform order of |
| operations and solve |
| equation. |

$13 = x$

Answer: $x = 13$

Example 2: In the right triangle shown, find x.

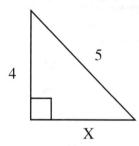

Solution:

$a^2 + b^2 = c^2$

$x^2 + 4^2 = 5^2$

| Write the formula. |
| Substitute. |

$x^2 + 16 = 25$

$x^2 + 16 - 16 = 25 - 16$

$\sqrt{x^2} = \sqrt{9}$

| Perform order of |
| operations and solve |
| equation. |

$x = 3$

Answer: $x = 3$

Write the length of the missing side in the space provided.

1.

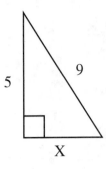

Answer _____

4.

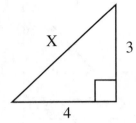

Answer _____

2.

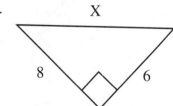

Answer _____

5.

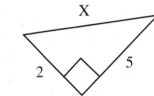

Answer _____

4.

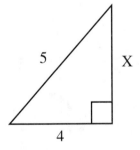

Answer _____

6.

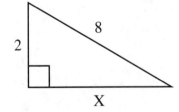

Answer _____

122

59. Sample Examinations and Answers

SAMPLE EXAM # 1

Write the letter of the correct answer in the space provided.

1. Write six million three hundred fifty four.
   (a) 600,354   (b) 6,000,354   (c) 60,000,354
   (d) 60,254   (e) 6,003,540   Answer _____

2. $671 - 89$
   (a) 760   (b) 572   (c) 682
   (d) 581   (e) 582   Answer _____

3. A play starts at 6:45 p.m. and ends at 9:30 p.m.
   How long is the play?
   (a) 3 hours 45 minutes   (b) 2 hours 30 minutes
   (c) 2 hours 45 minutes   (d) 1 hour 30 minutes
   (e) 4 hours 45 minutes   Answer _____

4. $\frac{2}{7} + \frac{1}{6}$

   (a) $\frac{19}{42}$ (b) $\frac{3}{13}$ (c) $\frac{3}{42}$
   (d) $\frac{1}{14}$ (e) $\frac{1}{7}$ Answer _____

5. $4.08 + .48 + 40$
   (a) 4.96   (b) 44.56   (c) .496
   (d) 4.456   (e) .4456   Answer _____

6. What is 30% of 90?
    (a) 27   (b) 270   (c) .27
    (d) 300   (e) 3,000   Answer _____

7. $4(-1)^2 - 3(7)$
    (a) -25 (b) 25   (c) 17
    (d) -17 (e) 5   Answer _____

8. Change $\frac{2}{9}$ to a decimal, rounded to the
    nearest hundredth.
    (a) .22   (b) .23   (c) .222
    (d) .02   (e) .022   Answer _____

9. Which number is the smallest?
    (a) .04   (b) .041   (c) .004
    (d) .0045   (e) .401   Answer _____

10. $12.2 - 3.49$
    (a) 15.69   (b) 1.569   (c) 8.71
    (d) 87.1   (e) .871   Answer _____

11. Which of the fractions is the smallest?
    (a) $\frac{5}{9}$ (b) $\frac{2}{7}$ (c) $\frac{3}{5}$
    (d) $\frac{1}{5}$ (e) $\frac{2}{9}$ Answer _____

12. $2424 \div 12$
    (a) 22   (b) 101   (c) 11
    (d) 202   (e) 200   Answer _____

13. $1\frac{3}{4} \div 7$

    (a) $\frac{1}{7}$ (b) $\frac{1}{4}$ (c) $\frac{2}{3}$

    (d) $\frac{1}{6}$ (e) $\frac{3}{7}$ Answer _____

14. Jose has a 92 average on three chemistry tests. On his first two tests he received a 97 and 91. What must his grade be on the third test?

    (a) 94   (b) 91   (c) 87

    (d) 88   (e) 89   Answer _____

15. If 20% of a number is 4, find the number.

    (a) .8   (b) 8   (c) 80

    (d) 2   (e) 20   Answer _____

16. $3\frac{5}{6} - 1\frac{1}{3}$

    (a) $2\frac{1}{2}$ (b) $3\frac{1}{2}$ (c) $5\frac{1}{6}$

    (d) $2\frac{1}{3}$ (e) $3\frac{2}{3}$ Answer _____

17. A $30 blouse is reduced by 10%. What is the sale price?

    (a) $3   (b) $24   (c) $33

    (d) $27   (e) $36   Answer _____

18. What is the cost to carpet a room 8 yards by 6 yards at $5 per square yard?

    (a) $400   (b) $300   (c) $240

    (d) $140   (e) $280   Answer _____

19. $3.1 \times .004$
    (a) .124   (b) 12.4   (c) 1.24
    (d) 124   (e) .0124   Answer _____

20. $.022 \div .02$ equals:
    (a) 1.1   (b) 11   (c) .11
    (d) .011   (e) 110   Answer _____

21. Add $2a^2 - 7a$ and $3 - a$.
    (a) $6a^2 - 7a + 3$ (b) $2a^3 - 7a - 3$
    (c) $2a^2 - 8a + 3$ (d) $6a^2 - 6a + 3$
    (e) $2a^3 + 8a - 3$ Answer _____

22. Simplify $\dfrac{12x^2 - 6x}{2x}$.

    (a) $6x - 3$ (b) $6x^2 - 3x$ (c) $6x^3 - 3x^2$
    (d) $6x^2 - 3$ (e) $6x^3 - 6x$ Answer _____

23. Find the value of "q" quarters and "n" nickels.
    (a) $q + n$ (b) $25(q + n)$ (c) qn
    (d) $5(q + n)$ (e) $25q + 5n$ Answer _____

24. Solve for x: $3x - 6 = x + 4$
    (a) -5 (b) 3   (c) -3
    (d) 5   (e) 2   Answer _____

25. Solve for x: $\frac{x}{2} + \frac{x}{4} = 3$

    (a) 3   (b) -3 (c) 4
    (d) -4 (e) 5   Answer _____

26. Simplify $(2x^3y^4)^3$
    (a) $8x^6y^7$ (b) $8x^9y^4$ (c) $6x^9y^{12}$
    (d) $8x^9y^{12}$ (e) $8x^6y^{12}$ Answer _____

27. If a = 3 and b = 4, find the value of $3a^2 - 2b^2$.
    (a) 5   (b) −59   (c) −5
    (d) 59   (e) 30   Answer _____

28. Factor $6x^2 - 3x$.
    (a) $3(2x^2 - x)$ (b) $3x(2x - 1)$ (c) $3x(2x - 3)$
    (d) $3(2x^2 - 3x)$ (e) $3(6x^2 - x)$ Answer _____

29. Find the product of $3x^2y^3$ and $-2xy$.
    (a) $6xy^2$ (b) $-6x^3y^4$ (c) $-6xy$
    (d) $6x^3y^4$ (e) $6x^2y$ Answer _____

30. Simplify $6a^2 - 2a(a + 3)$.
    (a) $4a^2 - 6a$ (b) $8a^2 - 6a$ (c) $2a^2 - 3a$
    (d) $4a^2 - 3a$ (e) $4a^2 + 6a$ Answer _____

31. Subtract $3x^2 - 6$ from $6x^2 - 7$.
    (a) $9x^2 - 1$ (b) $-3x^2 + 1$ (c) $3x^2 - 1$
    (d) $9x^2 + 1$ (e) $3x^4 - 1$ Answer _____

32. If $2x - y = 6$ and $4x + y = 6$ then,
    (a) x = 2, y = 1   (b) x = 2, y = −2
    (c) x = 2, y = 4   (d) x = 4, y = −1
    (e) x = 3, y = 3   Answer _____

33. In the following right triangle, find the value of x:

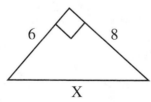

    (a) 4   (b) 100   (c) 8
    (d) 5   (e) 10   Answer _____

34. Which of the following points lies on the graph $2x + y = 6$?
    (a) (3,3)   (b) (−3,−3)   (c) (2, 2)
    (d) (−2,−2)   (e) (6, 1)   Answer _____

35. Solve for x: $2x + 3y = 7$
    (a) $7-3y$ (b) $7+3y$ (c) $\frac{7-2x}{3}$
    (d) $\frac{7-3y}{2}$ (e) $\frac{7+3y}{2}$ Answer _____

36. Write an equation for the line.

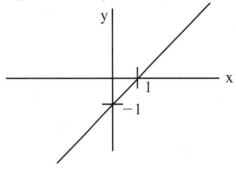

    (a) $x + 7 = 1$ (b) $x - y = 1$ (c) $x = y + 2$
    (d) $x + y = 2$ (e) $x + 2y = 2$ Answer _____

37. There are 80 mathematics professors at City College in which 25% are women. Without firing anyone, how many additional women must be hired in order to bring the percentage of women up to 50%?
    (a) 20   (b) 30   (c) 40
    (d) 50   (e) 60   Answer _____

38. Solve for x: $\frac{x}{2} - 1 = \frac{x + 3}{4}$
    (a) −7   (b) 4   (c) −4
    (d) 1   (e) 7   Answer _____

39. A toy store owner finds that 2 out of 7 toys sold are returned. If 350 toys are sold in a week, how many are expected to be returned?

(a) 50   (b) 100   (c) 150

(d) 200   (e) 250   Answer _____

40. If (o, b) lies on the graph, $y = 3x - 8$ then, $b =$

(a) 3   (b) -3 (c) 8

(d) 6   (e) -8 Answer _____

ANSWERS: SAMPLE EXAM # 1

1.	b	21.	c
2.	e	22.	a
3.	c	23.	e
4.	a	24.	d
5.	b	25.	c
6.	a	26.	d
7.	d	27.	c
8.	a	28.	b
9.	c	29.	b
10.	c	30.	a
11.	d	31.	c
12.	d	32.	b
13.	b	33.	e
14.	d	34.	c
15.	e	35.	d
16.	a	36.	b
17.	d	37.	c
18.	c	38.	e
19.	e	39.	b
20.	a	40.	e

SAMPLE EXAM # 2

B. Write the letter of the correct answer in the space provided.

1. Write three million forty five.
   (a) 300,045   (b) 3,000,045   (c) 3,000,450
   (d) 30,045   (e) 300,450   Answer _____

2. $421 - 37$
   (a) 484   (b) 458   (c) 358
   (d) 494   (e) 384   Answer _____

3. A movie starts at 6:15 p.m. and ends at 9:05 p.m.
   How long was the movie?
   (a) 3 hours 50 minutes   (b) 2 hours 45 minutes
   (c) 4 hours 20 minutes   (d) 4 hour 50 minutes
   (e) 2 hours 50 minutes   Answer _____

4. $\frac{4}{9} + \frac{1}{3}$

   (a) $\frac{5}{12}$ (b) $\frac{5}{9}$ (c) $\frac{4}{27}$
   (d) $\frac{7}{9}$ (e) $\frac{2}{3}$ Answer _____

5. $4.7 - 1.89$
   (a) 6.59   (b) 65.9   (c) 2.81
   (d) 28.1   (e) .281   Answer _____

6. Change $\frac{7}{9}$ to a decimal rounded to the nearest
   hundredth.
   (a) .78   (b) .79   (c) .777
   (d) .709   (e) .785   Answer _____

7. $6(-2)^2-3(6)$
   (a) 6   (b) -6 (c) 42
   (d) -4 (e) 12   Answer _____

8. What is 40% of 70?
   (a) .28   (b) 175   (c) 28
   (d) 1.75   (e) 1.19   Answer _____

9. Which fraction is the smallest?
   (a) $\frac{2}{7}$ (b) $\frac{1}{3}$ (c) $\frac{4}{9}$
   (d) $\frac{3}{8}$ (e) $\frac{2}{5}$ Answer _____

10. $\frac{2}{3} \div 6$

    (a) 4   (b) $\frac{1}{9}$ (c) $\frac{1}{4}$
    (d) $\frac{5}{9}$ (e) $\frac{2}{3}$ Answer _____

11. Zelma has a 94 average on three mathematics tests. On her first two tests, she received 89 and 95. What must her grade be on the third test?
    (a) 95   (b) 96   (c) 97
    (d) 98   (e) 99   Answer _____

12. If 40% of a number is 60, find the number.
    (a) 250   (b) 100   (c) 150
    (d) 240   (e) 24   Answer _____

13. Which number is the smallest?
    (a) .076   (b) .7   (c) .07
    (d) .071   (e) .075   Answer _____

14. $6.1+.61+61$
    (a) 7.32   (b) 6.771   (c) 73.2
    (d) .6771   (e) 67.71   Answer _____

15. Express 40% as a fraction?

    (a) $\frac{2}{5}$ (b) $\frac{7}{10}$ (c) $\frac{3}{5}$

    (d) $\frac{4}{5}$ (e) $\frac{3}{10}$ Answer _____

16. $4\frac{5}{9} - 2\frac{1}{3}$

    (a) $4\frac{2}{9}$ (b) $3\frac{2}{9}$ (c) $2\frac{2}{9}$

    (d) $2\frac{2}{3}$ (e) $3\frac{1}{3}$ Answer _____

17. If pencils sell for $.12 each, how many can be purchased for $24?

    (a) 48   (b) 480   (c) 20

    (d) 40   (e) 200   Answer _____

18. A $45 jacket is reduced by 20%.
What is the discount?

    (a) $9   (b) $36   (c) $25

    (d) $50   (e) $60   Answer _____

19. Find the cost to carpet a room 7 yards by 6 yards at $4 per square yard.

    (a) $28   (b) $24   (c) $42

    (d) $168   (e) $104   Answer _____

20. $3.6 \div 36$ equals:

    (a) 10   (b) 1   (c) .01

    (d) .1   (e) 100   Answer _____

21. Solve for x: $4x - 2 = x + 7$
     (a) -3　　　　(b) 2　　　　(c) -2
     (d) 4　　　　 (e) 3　　　　 Answer _____

22. Simplify $(3x^2y^3)^4$.
     (a) $12x^8y^{12}$　　(b) $81x^8y^{12}$　　(c) $12x^6y^7$
     (d) $81x^6y^7$　　　(e) $12x^6y^7$　　　 Answer _____

23. Find the value of "d" dimes and "n" nickels.
     (a) $d + n$　　　(b) $10(d + n)$　　(c) 5
     (d) $5(d + n)$　　(e) $10d + 5n$　　 Answer _____

24. Solve for x: $\frac{x}{3} + \frac{x}{5} = 8$

     (a) 15　　　　(b) -15　　　(c) 8
     (d) -8　　　　(e) 6　　　　 Answer _____

25. If $x = 4$ and $y = 5$, find the value of $3x^2 + 4y^2$.
     (a) 32　　　　(b) 27　　　　(c) 144
     (d) 29　　　　(e) 148　　　 Answer _____

26. Simplify $\frac{16x^2 - 32x}{8x}$

     (a) $2x - 4$　　(b) $2x + 4$　　　(c) $2x^3 - 4x^2$
     (d) $2x^2 - 4x$　(e) $2x^3 - 4x$　　 Answer _____

27. Factor $9y^2 - 6y$.
     (a) $3(3y^2 - 2y)$　(b) $y(9y - 6)$　　(c) $3y(3y - 2)$
     (d) $3y(7y - 6)$　(e) $y(3y - 6)$　　 Answer _____

28. Add $3x^2 - 6x$ and $4 - x$.
     (a) $3x^2 - 6x+4$　(b) $3x^2 - 7x+4$　(c) $3x^2 - 6x - 4$
     (d) $3x^2 - 3x$　　(e) $7x^2 - 7x$　　 Answer _____

29. Simplify $3x^2 - 2x(x + 3)$.
    (a) $x^2 - 6x$ (b) $x^2 + 6x$ (c) $5x^2 - 6x$
    (d) $5x^2 + 6x$ (e) $-x^2 - 6x$ Answer _____

30. Find the product of $4a^2b^3$ and $-3ab$.
    (a) $-12a^2b^3$ (b) $-12a^3b^4$ (c) $64a^2b^3$
    (d) $-64a^2b^3$ (e) $12a^2b^3$ Answer _____

31. If $3x - y = 4$ and $x + y = 8$, then,
    (a) $x = 5, y = 2$ (b) $x = 3, y = 5$
    (c) $x = 2, y = 3$ (d) $x = 4, y = 1$
    (e) $x = 6, y = 2$ Answer _____

32. Subtract $3a^2 - 6b^2$ from $6b^2 - 3a^2$.
    (a) 0 (b) $-12b^2 - 6a^2$ (c) $3a^2 - 12b^2$
    (d) $9a^2 - 9b^2$ (e) $-6a^2 + 12b^2$ Answer _____

33. In the following right triangle, find the value of x:

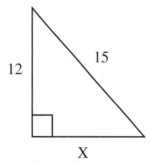

    (a) 9   (b) 369   (c) 11
    (d) 15   (e) 17   Answer _____

34. Which of the following points lies on the graph $3x + 2y = 5$:
    (a) $(-1, -1)$ (b) $(2, -4)$ (c) $(-2, 4)$
    (d) $(3, 7)$ (e) $(1, 1)$ Answer _____

35. Solve for y: $2x - 4y = 14$.
    (a) $7 + x$ (b) $\frac{7 - x}{2}$ (c) $-7 - x$
    (d) $\frac{7 + x}{2}$ (e) $\frac{-7 + x}{2}$ Answer _____

36. Write an equation for the line.

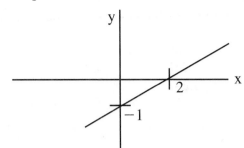

(a) $x - 2y = 2$ (b) $x - y = 2$ (c) $x - 2y = -2$
(d) $x - 2y = 1$ (e) $x - y = -1$ Answer _____

37. Solve for x: $\frac{x}{3} - 2 = \frac{x + 6}{6}$.

(a) 18   (b) -18 (c) 9
(d) -9 (e) 8   Answer _____

38. Martin Luther King High School has 60 psychologists
in which 20% are women. Without firing anyone,
how many additional women must be hired in order
to bring the percentage of women up to 50%?
(a) 24   (b) 35   (c) 12
(d) 25   (e) 36   Answer _____

39. A secretary earns $60 in 8 hours. At this rate of pay,
how much will she earn in 20 hours?
(a) $100   (b) $150   (c) $200
(d) $75   (e) $50   Answer _____

40. If $a = 3bc^2$, find "a" when b = 4 and c = 5.
(a) 240   (b) 120   (c) 70
(d) 60   (e) 300   Answer _____

ANSWERS: SAMPLE EXAM #2

1.	b	21.	e
2.	e	22.	b
3.	e	23.	e
4.	d	24.	a
5.	c	25.	e
6.	a	26.	a
7.	a	27.	c
8.	c	28.	b
9.	a	29.	a
10.	b	30.	b
11.	d	31.	b
12.	c	32.	e
13.	c	33.	a
14.	e	34.	e
15.	a	35.	e
16.	c	36.	a
17.	e	37.	a
18.	a	38.	e
19.	d	39.	b
20.	d	40.	e

60. Answer Key

PAGE 3
1. 9,826
2. 38,677
3. 7,134
4. 11
5. 2
6. 661
7. 33,558
8. 26,325
9. 73,290
10. 101
11. 39
12. 560
13. 21,528
14. 55
15. 6,033
16. 101
17. 3,648
18. 99
19. 4,347
20. 3,880
21. 3,135
22. 228
23. 56
24. 71,063
25. 119

PAGES 4 & 5
1. C
2. E
3. A
4. B
5. D
6. A
7. B
8. E
9. B

PAGES 6 & 7
1. 9 hrs. 6 mins.
2. 6 yds. 1 ft. 4 in.
3. 1 hr. 57 mins.
4. 7 in.
5. 1 hr. 55 mins.
6. 1 ft. 7 in.
7. 2 hrs. 40 mins.
8. 3 hrs. 47 mins.
9. 37 mins.
10. 1 hr. 5 mins.

PAGES 7 & 8
1. 36.7
2. 60
3. 80
4. 14
5. 20
6. 90
7. 100
8. 12
9. 40
10. 10

PAGES 9 & 10
1. $\frac{1}{3}$
2. $1\frac{1}{4}$
3. $\frac{3}{4}$
4. $1\frac{1}{3}$
5. $4\frac{1}{2}$
6. $\frac{2}{9}$
7. $\frac{2}{3}$
8. $1\frac{1}{2}$
9. $4\frac{1}{4}$
10. $3\frac{1}{2}$
11. $\frac{2}{3}$
12. $\frac{3}{14}$

Answer Key

PAGES 11 & 12

1. $\frac{19}{42}$
2. $\frac{7}{8}$
3. $7\frac{1}{9}$
4. $1\frac{5}{12}$
5. 5
6. $5\frac{7}{8}$
7. $1\frac{2}{35}$
8. $\frac{19}{45}$
9. $\frac{11}{12}$
10. $8\frac{1}{4}$
11. $\frac{46}{63}$
12. 8
13. $4\frac{9}{40}$
14. $8\frac{19}{28}$

PAGE 14

1. $\frac{1}{4}$
2. $\frac{5}{8}$
3. $2\frac{5}{42}$
4. $3\frac{7}{36}$
5. $5\frac{1}{8}$
6. $\frac{2}{9}$
7. $2\frac{4}{5}$
8. $\frac{3}{20}$
9. $\frac{1}{2}$
10. $\frac{1}{21}$
11. $1\frac{7}{9}$
12. $2\frac{47}{56}$
13. $8\frac{1}{3}$
14. $\frac{7}{15}$
15. $1\frac{27}{28}$
16. $1\frac{23}{56}$

PAGE 16

1. $\frac{1}{21}$
2. $\frac{1}{12}$
3. $\frac{6}{7}$
4. 6
5. 2
6. $12\frac{1}{7}$
7. $3\frac{1}{2}$
8. $\frac{3}{35}$
9. $\frac{2}{3}$
10. $\frac{1}{14}$
11. $\frac{1}{4}$
12. 15
13. 14
14. $11\frac{2}{3}$
15. 1
16. $\frac{2}{15}$

PAGE 18

1. $\frac{1}{2}$
2. $\frac{7}{8}$
3. 48
4. $\frac{1}{14}$
5. $1\frac{1}{5}$
6. $\frac{1}{3}$
7. $\frac{2}{3}$
8. 4
9. 3
10. 2
11. 21
12. $\frac{1}{18}$
13. $1\frac{1}{3}$
14. 34
15. $4\frac{2}{9}$
16. $\frac{2}{3}$

PAGES 21 & 22: PART A

1. E
2. A
3. C
4. D
5. E

PAGES 22 & 23: PART B

1. C
2. E
3. A
4. E
5. A
6. C
7. B
8. A
9. E

PAGE 25: PART A

1. E
2. B
3. B
4. A
5. E
6. C

PAGE 26: PART B

1. B
2. D
3. E
4. E
5. C
6. A

PAGE 27: PART A
1. 29.981
2. 41.996
3. 17.802
4. 9.865
5. 23.45
6. 8.325
7. 97.989
8. 27.84
9. 79.92
10. 79.665

PAGES 27 & 28: PART B
1. 25.1
2. 9.35
3. 2.922
4. 12.75
5. 3.89
6. 71.28
7. 1.3
8. 5.911
9. 235.22
10. 548.33

PAGES 28 & 29: PART A
1. .20
2. .234
3. .96
4. .017
5. .081
6. 2109.12
7. .0032
8. 14.4
9. 1.26
10. .0048
11. 82.62
12. 5152.17

PAGE 29: PART B
1. $2.94
2. $26.40
3. $3.95
4. $205.80
5. $45.00

PAGE 30: PART A
1. 60
2. 2,400
3. 200
4. .07
5. 5
6. 12,000
7. .14
8. 12
9. 50
10. 1.01
11. 200
12. .6

PAGE 30: PART B
1. 40
2. 24
3. 20
4. 16

PAGE 31: PART A
1. .24
2. .65
3. .045
4. .03
5. 4.5
6. .35
7. .36
8. 1.75

PAGES 32: PART B
1. $\frac{7}{20}$
2. $\frac{1}{50}$
3. $\frac{1}{40}$
4. $1\frac{1}{5}$
5. $4\frac{7}{20}$
6. $\frac{9}{200}$
7. $\frac{4}{25}$
8. $6\frac{2}{5}$
9. $\frac{89}{1000}$
10. $\frac{29}{40}$

PAGES 32 & 33: PART C
1. $\frac{7}{20}$
2. $\frac{2}{25}$
3. $\frac{19}{100}$
4. $\frac{6}{25}$
5. $1\frac{1}{2}$
6. $\frac{7}{10}$
7. $\frac{79}{100}$
8. $2\frac{9}{20}$
9. $\frac{9}{25}$
10. $\frac{1}{25}$

PAGE 33: PART D
1. .39
2. .07
3. 1.25
4. .008
5. .06
6. .54
7. 4.29
8. .007
9. .04
10. .16

PAGE 35: PART A
1. .38
2. .22
3. .86
4. .21
5. .29
6. .18
7. .89
8. .83

PAGE 35: PART B
1. .273
2. .769
3. .286
4. .375
5. .471
6. .579
7. .571
8. .778

PAGES 36 & 37
1. $5.25
2. $4.12
3. $111.00
4. $170.00
5. $1,003.00
6. $3,675.00
7. $3.44
8. $1,820.00
9. $10.65
10. $3.70

PAGES 38 & 39
1. C
2. A
3. E
4. D
5. A
6. D
7. A
8. B

PAGE 40
1. $97.20
2. $30,740.00
3. $38.00
4. $45.00
5. $8,320,000.00
6. $16.80
7. $20.87

PAGES 41 & 42
1. $386.75
2. $71.20
3. 7,440,000
4. $22,000.00
5. $34.91
6. $16.25
7. $1.68
8. 1,980,000

PAGE 44
1. 40 sq. yds
2. 1,950 sq. in.
3. $504.00
4. $315.00
5. 30 ft.
6. 110 ft.
7. $840.00
8. $378.00
9. $864.00
10. 12 in.

PAGE 55: PART A
1. −7
2. 5
3. 3
4. 3
5. −11
6. −1
7. 0
8. −3

PAGE 56: PART B
1. 4
2. 4
3. 5
4. −3
5. 9
6. −4
7. 14
8. −2
9. −11
10. 1
11. 5
12. 11

PAGES 58 & 59
1. -12
2. -35
3. 12
4. 15
5. -56
6. 60
7. 9
8. -3
9. -4
10. 3
11. -7
12. -9

PAGE 61
1. -9
2. 9
3. 48
4. 16
5. -8
6. -24
7. -7
8. 30
9. 31
10. 7
11. 144
12. -243
13. 78
14. -169
15. -48
16. 42
17. 165
18. -11
19. 81
20. -32

PAGES 62 & 63: PART A
1. $5 + x$
2. $x - 7$
3. $x - 6$
4. $x + 5$
5. $7x$
6. $x - 4$
7. $x + 14$
8. $4 + \frac{x}{2}$
9. $3 + 4x$
10. $3x - 6$

PAGE 63: PART B
1. $5n + 25q$
2. $10d + 25q$
3. $10d + 5n$
4. $5n + 10d + 25q$

PAGE 65
1. $9x + 15y$
2. $7a - 9b$
3. $8x^3 + 6x^2 - 2x$
4. $10b + 5$
5. $4x^2 - 10x + 8$
6. $10y^2 + 10y - 8$
7. $-4a^2 - 3a - 2$
8. $4x - 14$
9. $8x^2 - 16x$
10. $3x^2 + 8x - 7$
11. $-14y$
12. $8x^2 - 11x + 16$
13. $-2x - 3$
14. $-6x^2 + 14x - 12$
15. $8x^2 + 3x - 12$
16. $-3x^2 + 3x - 3$
17 $2x^2 - 10x - 6$
18. $-6x^2$

PAGES 66 & 67
1. x^5
2. x^2y^3
3. a^3
4. a^9
5. 2^{11}
6. a^4
7. y^9
8. y^8x
9. y^3x
10. y^9

PAGES 67 & 68
1. $12x^3y^3$
2. $-14x^5y^8$
3. $-14r^2s^2t^2$
4. $-12x^5y^{11}$
5. $-12a^2b^3c^2$
6. $32x^5$
7. $-14x^2y^3z^2$
8. $-12a^3b^3c^3$
9. $-14r^3s^2t^3$
10. $12x^4y^4z^3$

PAGES 68 & 69

1. $3X-6$
2. $-2y+8$
3. a^2+4a
4. $-b^2+7b$
5. $-3x^2+18x$
6. $-4y^2+28y$
7. $-10z^3+35z^2$
8. $-56x^3-64x^2$
9. $-10s^3+35s^2$
10. $8a^2b^2-28ab^2$
11. $21x^3y^3-7xy^2$
12. $-12a^3b^3-12a^2b^4$
13. $-28a^3b^3+21a^4b^2$
14. $-6x^4y^6+x^3y^5$
15. $-21x^4y+63x^2y^2$
16. $12x^3y^2+15x^2y^3$
17. $-4x^2+28x$
18. $18x^3-21x^2$

PAGE 70

1. x^2+2x-3
2. x^2+x-20
3. $4x^2+2x-12$
4. $6x^2-x-2$
5. y^2-4
6. y^2-25
7. $x^2-4x-21$
8. x^2+x-30
9. $9x^2-36$
10. $6x^2+4x-32$
11. x^2-4x+4
12. $X^2+10x+25$

PAGE 71

1. $x^3+5x^2+10x+8$
2. $y^3-2y^2-2y-24$
3. $2x^3-3x^2+6x+4$
4. y^3+27
5. $x^3+2x^2-2x-12$
6. $y^3-4y^2+7y-12$
7. $5y^4-13y^3+21y^2+4y+10$
8. $2x^4-7x^3+3x^2-x+3$
9. $y^3-2y^2-8y-35$
10. $3x^3-18x^2+45x-42$

PAGES 72 & 73

1. -10
2. -12
3. -25
4. 14
5. -53
6. -7
7. 166
8. 33
9. -60
10. 48
11. -48
12. -8
13. 46
14. -100

PAGES 74

1. y^2+10y
2. x^2+6x
3. x^3-7x^2+3x
4. $-4y^3+36y^2+4y$
5. $-x^2+28x$
6. $-y^2+4y$
7. $-2x^3-x^2$
8. $-4y^3-25y^2$
9. a^2b+14a
10. $-7x^3-20x^2y$
11. $52x^2y^2-7x^2y$
12. $6x^2y+15xy^2-6xy$
13. $-15a^2b+6ab^2$
14. $10x^2y+15xy^2$
15. y^2+14y
16. x^2-4x
17. $-4x^3-21x^2$
18. $5y^3+28y^2$
19. x^2-8x
20. $6y^3+40y^2$

PAGES 75 & 76

1. C
2. B
3. A
4. E
5. E
6. A
7. E
8. B
9. A

PAGE 77

1. x^2
2. a^5
3. b^6
4. $\frac{x7}{y}$
5. x^{-4}
6. a^5
7. b^4
8. $\frac{x^8}{y^2}$

PAGE 78

1. $3x$
2. 4
3. $-2a+4$
4. $-2y+4$
5. $-2a+1$
6. $-2y^2 + \frac{3}{2}y$

PAGES 79 & 80

1. $3(x^2+2)$
2. $12(a-2)$
3. $2(3x^2-8)$
4. $x(x+6)$
5. $a^3(a+7)$
6. $6x(x+2)$
7. $3x(x-2)$
8. $9(x^2-2)$
9. $3x(xy+2)$
10. $10y(x^2+2y)$
11. $9x^6y(2x-y^5)$
12. $6x^2y^6(xy-2)$
13. $x(x-4)$
14. $3(2a^2+b^2)$
15. $6ab(1-2ab)$
16. $7x^2y^3(y-2x^{12})$
17. $9(x^2+2y^2)$
18. $7x(x+2y)$

PAGE 81

1. $(x+4)(x+1)$
2. $(x+4)(x+3)$
3. $(x-2)(x-1)$
4. $(x-5)(x+1)$
5. $(y+4)(y-3)$
6. $(y-5)(y-4)$
7. $(x+6)(x+1)$
8. $(x+7)(x+6)$
9. $(y+6)(y-3)$
10. $(x+8)(x+7)$
11. $2(x+2)(x+1)$
12. $2y(y^2+2y+15)$
13. $2(y+7)(y+2)$
14. $5x(x+5)(x+4)$

PAGES 82 & 83

1. $(x+5)(x-5)$
2. $(y+8)(y-8)$
3. $4(x^2+4)$
4. $(x+7)(x-7)$
5. $(3x+4y)(3x-4y)$
6. $(7y+5x)(7y-5x)$
7. $3(x+4)(x-4)$
8. $2(x+3)(x-3)$
9. $(3+x)(3-x)$
10. $(4+y)(4-y)$
11. $(x+4)(x-4)$
12. $(x^3+6)(x^3-6)$
13. $(x^4+9)(x^2+3)(x^2-3)$
14. $9(x^2+3)(x^2-3)$

PAGES 84 & 85

1. 7
2. -3
3. 3
4. -3
5. 8
6. -5
7. -2
8. 4
9. 2
10. 8
11. -3

PAGE 86

1. -4
2. -23
3. 4
4. 9
5. 60
6. -12
7. 20
8. -28
9. 6

PAGE 88
1. $x = 7+y$
2. $y = x-3$
3. $a = 7b$
4. $b = \frac{a}{2}$
5. $x = \frac{y+2}{3}$
6. $y = \frac{4x-6}{5}$
7. $x = 2y-8$
8. $x = 3y+21$
9. $z = 6-x-y$
10. $y = \frac{3x+7y-9}{4}$
11. $a = 4c+b$
12. $x = 2-y$
13. $x = y-z-6$

PAGE 90
1. $x = 3$
2. $x = \frac{1}{2}$
3. $x = 6, x-4 = 2$
4. $x = -2$
5. $x = 7$
6. $x = 9$
7. $x = 2$
8. $x = 3$

PAGE 92
1. $x = 2$
2. $x = 6$
3. $x = 240$
4. $x = 4$
5. $x = \frac{-4}{5}$
6. $x = \frac{-1}{9}$

PAGE 93
1. 91 grams
2. 5 inches
3. 2,275 miles
4. 756 miles
5. 200 grams
6. 10 throws

PAGE 95
1. $x = 3, y = 1$
2. $a = 4, b = -1$
3. $x = 4, b = -\frac{1}{2}$
4. $x = -1, y = 4$
5. $a = 4, b = -2$

PAGE 96
1. $x = 4, y = 0$
2. $x = 2\frac{3}{5}, y = \frac{1}{5}$
3. $x = 3, y = 6$
4. $a = 0, b = 1$
5. $x = -2, y = 1$

PAGE 98
1. $x = \frac{10}{7}, y = \frac{3}{7}$
2. $x = 1, y = 1$
3. $x = -1, y = 3$
4. $x = 3, y = 0$
5. $x = 4, y = -1$
6. $x = -2, y = 4$
7. $x = 2, y = 2$

PAGE 99
1. $(-5, -2)$
2. $(-11, -1)$
3. $(-4, -1)$
4. $(5, 4)$
5. $(1, -3)$
6. $(4, -4)$
7. $(3, -2)$
8. $(5, -5)$
9. $(4, 3)$
10. $(-3, -1)$
11. $(-3, -2)$
12. $(-5, -3)$

PAGES 101 & 102
1. D
2. B
3. C
4. B
5. C
6. B
7. B

PAGE 103
1. $x = 3$
2. $x = 3$
3. $x = -2$
4. $x = 4$
5. $x = 3$
6. $y = -2$
7. $x = \frac{1}{3}$
8. $y = \frac{1}{2}$

PAGE 105

1a. $x - y = 2$

1b. $-4x + 3y = 12$

2a. $-2x + y = 2$

2b. $x + y = 2$

PAGE 107

1. $\frac{3}{2}$

2. $\frac{4}{3}$

3. $\frac{-1}{4}$

4. 1

5. $\frac{5}{3}$

6. 3

7. 1

8. $\frac{5}{4}$

9. 3

10. $\frac{-2}{3}$

PAGE 108

1. $y = 4x + 1$

2. $y = 3x - 6$

3. $y = -2x + 4$

4. $y = -5x - 6$

5. $y = -x - 1$

6. $y = 2x + 2$

7. $y = \frac{2}{3}x + 6$

8. $y = \frac{-2}{5}x + \frac{3}{4}$

PAGE 109

1. $m = -4, b = 6$

2. $m = -3, b = 4$

3. $m = 2, b = -5$

4. $m = 5, b = -6$

5. $m = -2, b = 3$

6. $m = \frac{-1}{2}, b = \frac{3}{2}$

7. $m = 1, b = -2$

8. $m = \frac{-2}{3}, b = 2$

PAGE 113

1. $61°$

2. $121°$

3. $150°$

4. $18°, 72°$

5. $60°, 120°$

6. $39°$

7. $120°$

8. $60°, 120°$

PAGE 115

1. $55°$

2. $90°$

3. $50°$

4. $90°$

PAGE 116

1. 88 cm.

2. 1.75 cm.

3. 154 sq. cm.

4. 44 cm.

5. 3.5 cm.

6. 201 cm.

PAGE 120

1. 52 cm.

2. 12 cm.

3. 4 cm.

4. 8 cm.

5. 36 sq. cm.

6. 4 cm.

PAGE 122

1. $\sqrt{56}$

2. 10

3. 3

4. 5

5. $\sqrt{29}$

6. $\sqrt{60}$

61. Homework Section

Assignment: BASIC SKILLS

Exercise 1:
Book pages 1–3

1. $2,049 + 35 + 6,849$ Answer _____

2. $34,591 + 67 + 1,201$ Answer _____

3. $367 + 129 + 1,750$ Answer _____

4. $804 - 309$ Answer _____

5. $4,001 - 3,999$ Answer _____

6. 429×202 Answer _____

7. 205×310 Answer _____

8. $2,424 \div 12$ Answer _____

9. $3,075 \div 25$ Answer _____

10. $405 + 78 + 3,619$ Answer _____

11. $604 - 409$ Answer _____

12. 341×209 Answer _____

13. $3,090 \div 41$ Answer _____

14. $607 - 39$ Answer _____

15. $307 + 27 + 1,707$ Answer _____

16. 200×300 Answer _____

17. $1,290 - 748$ Answer _____

18. $4,000 \times 3,000$ Answer _____

19. $2,037 + 145 + 89$ Answer _____

20. 307×125 Answer _____

21. $607 - 309$ Answer _____

22. $4,107 \div 34$ Answer _____

23. $741 - 59$ Answer _____

24. 210×16 Answer _____

25. $3,070 \div 45$ Answer _____

Exercise 2:
Book pages 4–5

Write in Numerals:

1. three hundred forty-nine   Answer _____

2. four hundred seven   Answer _____

3. two thousand three hundred forty-seven   Answer _____

4. eight thousand fifty five   Answer _____

5. twenty-five thousand three hundred   Answer _____

6. four-thousand one hundred five   Answer _____

7. sixty-seven thousand fifty-nine   Answer _____

8. thirty thousand one   Answer _____

9. one hundred eighty-nine thousand three hundred   Answer _____

10. seven hundred five thousand one hundred forty-nine   Answer _____

11. eight million seventy-nine thousand   Answer _____

12. six hundred nine   Answer _____

13. seven million eighty-six thousand seven   Answer _____

14. three hundred thousand four hundred fifty-six   Answer _____

15. nine million eight hundred seven thousand four   Answer _____

Exercise 3:
Book pages 5–7

Adding and subtracting units:

1. 5 hours 36 minutes
    +4 hours 39 minutes   Answer_____

2. 3 hours 4 minutes
    +7 hours 12 minutes   Answer_____

3. 7 hours 6 minutes
    −3 hours 10 minutes   Answer_____

4. 3 hours 24 minutes
    −1 hour 39 minutes   Answer_____

5. six pounds 15 ounces
    +five pounds 12 ounces   Answer_____

6. 12 pounds 13 ounces
    + 6 pounds 7 ounces   Answer_____

7. 15 pounds 5 ounces
    −12 pounds 13 ounces   Answer_____

8. 7 pounds 5 ounces
    − 4 pounds 15 ounces   Answer_____

9. A movie began at 5:04 p.m. and ended at 7:00 p.m.
    How long did it last?   Answer_____

10. Richard leaves his father's house at 4:45 p.m. and
    arrives home at 6:07 p.m. How long did he travel?   Answer_____

11. Seth buys a piece of wood 12 feet six inches long.
    He cuts a piece 3 feet 9 inches. How much wood is left?   Answer_____

12. Philip buys three pounds seven ounces of sweet potatoes
    and two pounds twelve ounces baking potatoes. How
    many pounds of potatoes did he buy?   Answer_____

Exercise 4:
Book pages 7–8

Average Problems:

1. 12, 14, 16   Answer_____

2. 39, 43, 44   Answer_____

3. 93, 106, 113   Answer_____

4. 70, 78, 83, 97   Answer_____

5. 47, 52, 54, 61, 71   Answer_____

6. 38, 50, 61, 62, 69   Answer_____

7. Zelma scores 97, 91, 94 on her Chemistry exams.
   Find her average.   Answer_____

8. Elliot has grades of 97, 100, 91 and 92 on his math
   quizzes. What is the average of these grades?   Answer_____

9. A salesman spends $48, $54, $49 and $57 per night
   for his hotel room. What is the average cost per night
   for the hotel room?   Answer_____

10. The average monthly snow for Buffalo is as follows:
    December: 2 feet 6 inches; January: 3 feet 4 inches;
    February: 1 foot 5 inches. What is the average monthly
    snow in Buffalo during this period?   Answer_____

11. A student scored 60, 0, 90 on her physics tests.
    What is the average?   Answer_____

12. The cost to produce four coats are $75, $85, $90 and $70.
    What is the average?   Answer_____

Exercise 5:
Book pages 9–10

Reducing Fractions:

1. $\frac{6}{8}$ Answer_____

2. $\frac{8}{6}$ Answer_____

3. $\frac{6}{32}$ Answer_____

4. $\frac{32}{6}$ Answer_____

5. $\frac{12}{34}$ Answer_____

6. $\frac{34}{12}$ Answer_____

7. $\frac{3}{9}$ Answer_____

8. $\frac{12}{9}$ Answer_____

9. $\frac{12}{16}$ Answer_____

10. $\frac{16}{12}$ Answer_____

11. $\frac{12}{38}$ Answer_____

12. $\frac{38}{12}$ Answer_____

Exercise 6:
Book pages 10–12

Addition of fractions:

1. $\frac{3}{8} + \frac{1}{4}$ Answer_____

2. $\frac{1}{3} + \frac{5}{9}$ Answer_____

3. $\frac{1}{2} + \frac{3}{4}$ Answer_____

4. $\frac{3}{7} + \frac{1}{5}$ Answer_____

5. $\frac{2}{5} + \frac{7}{8}$ Answer_____

6. $\frac{3}{7} + \frac{1}{6}$ Answer_____

7. $4 + \frac{1}{5}$ Answer_____

8. $6 + \frac{7}{9}$ Answer_____

9. $2\frac{3}{4} + 3\frac{5}{8}$ Answer_____

10. $5\frac{1}{4} + 1\frac{3}{7}$ Answer_____

11. $2\frac{1}{9} + 3\frac{5}{6}$ Answer_____

12. $3\frac{3}{4} + 4\frac{1}{4}$ Answer_____

13. $6 + \frac{1}{5}$ Answer_____

14. $\frac{5}{12} + \frac{5}{6}$ Answer_____

15. $2\frac{1}{7} + 3\frac{1}{4}$ Answer_____

16. $2\frac{1}{9} + 3\frac{5}{6}$ Answer_____

Exercise 7:
Book pages 12–14

Subtraction of fractions:

1. $\frac{3}{4} - \frac{1}{4}$ Answer_____
2. $\frac{7}{8} - \frac{1}{8}$ Answer_____
3. $\frac{3}{4} - \frac{3}{8}$ Answer_____
4. $\frac{7}{16} - \frac{1}{4}$ Answer_____
5. $\frac{3}{7} - \frac{1}{6}$ Answer_____
6. $\frac{7}{9} - \frac{3}{8}$ Answer_____
7. $\frac{3}{5} - \frac{1}{6}$ Answer_____
8. $\frac{4}{9} - \frac{3}{8}$ Answer_____

9. $3 - \frac{1}{4}$ Answer_____
10. $4 - \frac{4}{5}$ Answer_____
11. $6\frac{4}{9} - 2\frac{7}{9}$ Answer_____
12. $3\frac{2}{7} - 1\frac{4}{7}$ Answer_____
13. $5\frac{3}{4} - 2\frac{1}{8}$ Answer_____
14. $6\frac{7}{9} - 2\frac{1}{4}$ Answer_____
15. $5\frac{1}{6} - 2\frac{3}{7}$ Answer_____
16. $6\frac{3}{8} - 4\frac{6}{7}$ Answer_____

Exercise 8:
Book pages 14–16

Multiplication of fractions:

1. $\frac{3}{7} \times \frac{1}{9}$ Answer_____
2. $\frac{1}{4} \times \frac{8}{9}$ Answer_____
3. $\frac{3}{5} \times \frac{7}{9}$ Answer_____
4. $\frac{6}{7} \times \frac{1}{2}$ Answer_____
5. $\frac{3}{4} \times 16$ Answer_____
6. $\frac{3}{10} \times 50$ Answer_____
7. $4 \times \frac{1}{8}$ Answer_____
8. $12 \times \frac{1}{16}$ Answer_____

9. $3\frac{1}{4} \times 8$ Answer_____
10. $2\frac{1}{7} \times 49$ Answer_____
11. $2\frac{2}{3} \times 1\frac{4}{9}$ Answer_____
12. $3\frac{1}{5} \times 4\frac{3}{7}$ Answer_____
13. $3\frac{1}{8} \times 4$ Answer_____
14. $\frac{1}{5} \times 25$ Answer_____
15. $3\frac{2}{7} \times 1\frac{5}{6}$ Answer_____
16. $6\frac{7}{8} \times 1\frac{1}{9}$ Answer_____

Exercise 9:
Book pages 16–18

Division of fractions:

1. $\frac{5}{6} \div \frac{1}{6}$ Answer_____

2. $\frac{5}{9} \div \frac{5}{27}$ Answer_____

3. $\frac{3}{5} \div \frac{9}{25}$ Answer_____

4. $\frac{3}{8} \div \frac{5}{16}$ Answer_____

5. $3 \div \frac{1}{12}$ Answer_____

6. $4 \div \frac{1}{8}$ Answer_____

7. $\frac{3}{7} \div 5$ Answer_____

8. $\frac{4}{5} \div 12$ Answer_____

9. $1\frac{3}{4} \div 7$ Answer_____

10. $3\frac{2}{7} \div 1\frac{1}{14}$ Answer_____

11. $3\frac{1}{4} \div 4\frac{5}{8}$ Answer_____

12. $2\frac{5}{8} \div \frac{9}{16}$ Answer_____

13. $4\frac{3}{5} \div 1\frac{3}{10}$ Answer_____

14. $2\frac{1}{6} \div 3\frac{7}{12}$ Answer_____

15. $6 \div \frac{1}{12}$ Answer_____

16. $\frac{3}{7} \div 18$ Answer_____

Exercise 10:
Book pages 18–21

Which fraction is the smallest?

1. a. $\frac{2}{7}$ b. $\frac{3}{4}$ c. $\frac{1}{5}$ d. $\frac{2}{9}$ e. $\frac{3}{11}$ Answer_____

2. a. $\frac{3}{8}$ b. $\frac{1}{4}$ c. $\frac{2}{9}$ d. $\frac{4}{5}$ e. $\frac{2}{7}$ Answer_____

3. a. $\frac{1}{3}$ b. $\frac{2}{7}$ c. $\frac{3}{13}$ d. $\frac{1}{11}$ e. $\frac{2}{19}$ Answer_____

4. a. $\frac{2}{5}$ b. $\frac{2}{3}$ c. $\frac{3}{5}$ d. $\frac{7}{10}$ e. $\frac{4}{9}$ Answer_____

5. a. $\frac{9}{11}$ b. $\frac{7}{9}$ c. $\frac{10}{11}$ d. $\frac{7}{8}$ e. $\frac{19}{21}$ Answer_____

6. a. $\frac{1}{7}$ b. $\frac{1}{4}$ c. $\frac{3}{11}$ d. $\frac{3}{10}$ e. $\frac{2}{9}$ Answer_____

7. a. $\frac{5}{12}$ b. $\frac{7}{17}$ c. $\frac{1}{3}$ d. $\frac{2}{7}$ e. $\frac{3}{8}$ Answer_____

8. a. $\frac{2}{9}$ b. $\frac{1}{4}$ c. $\frac{7}{8}$ d. $\frac{2}{5}$ e. $\frac{3}{11}$ Answer_____

9. a. $\frac{7}{20}$ b. $\frac{1}{3}$ c. $\frac{5}{12}$ d. $\frac{5}{11}$ e. $\frac{4}{15}$ Answer_____

10. a. $\frac{8}{13}$ b. $\frac{7}{11}$ c. $\frac{5}{7}$ d. $\frac{13}{20}$ e. $\frac{7}{20}$ Answer_____

Exercise 11:
Book pages 22–23

Which fraction is the largest?

1.   a. $\frac{2}{5}$ b. $\frac{3}{4}$ c. $\frac{4}{11}$ d. $\frac{13}{16}$ e. $\frac{5}{12}$ Answer_____

2.   a. $\frac{7}{9}$ b. $\frac{1}{3}$ c. $\frac{8}{11}$ d. $\frac{3}{8}$ e. $\frac{7}{10}$ Answer_____

3.   a. $\frac{7}{15}$ b. $\frac{5}{7}$ c. $\frac{1}{2}$ d. $\frac{4}{5}$ e. $\frac{4}{7}$ Answer_____

4.   a. $\frac{1}{2}$ b. $\frac{2}{9}$ c. $\frac{1}{6}$ d. $\frac{2}{3}$ e. $\frac{13}{17}$ Answer_____

5.   a. $\frac{3}{5}$ b. $\frac{2}{3}$ c. $\frac{7}{12}$ d. $\frac{11}{15}$ e. $\frac{2}{7}$ Answer_____

6.   a. $\frac{3}{7}$ b. $\frac{3}{5}$ c. $\frac{2}{5}$ d. $\frac{5}{9}$ e. $\frac{7}{11}$ Answer_____

7.   a. $\frac{1}{8}$ b. $\frac{3}{4}$ c. $\frac{7}{9}$ d. $\frac{4}{5}$ e. $\frac{2}{7}$ Answer_____

8.   a. $\frac{1}{4}$ b. $\frac{3}{7}$ c. $\frac{4}{13}$ d. $\frac{2}{11}$ e. $\frac{4}{7}$ Answer_____

9.   a. $\frac{3}{4}$ b. $\frac{3}{7}$ c. $\frac{4}{13}$ d. $\frac{7}{9}$ e. $\frac{7}{8}$ Answer_____

10.   a. $\frac{5}{11}$ b. $\frac{7}{9}$ c. $\frac{10}{17}$ d. $\frac{5}{8}$ e. $\frac{19}{23}$ Answer_____

Exercise 12:
Book pages 24–26

A. Which number is the smallest?

1. (a) .41   (b) .44   (c) .46   (d) .49   (e) .40   Answer_____

2. (a) .52   (b) .5   (c) .55   (d) .59   (e) .50   Answer_____

3. (a) .256   (b) .251   (c) .252   (d) .257   (e) .258   Answer_____

4. (a) 2.85   (b) 2.75   (c) 2.764   (d) 2.759   (e) 2.77   Answer_____

5. (a) .09   (b) .009   (c) .0009   (d) .9   (e) .00009   Answer_____

6. (a) 7.45   (b) 7.451   (c) 7.5   (d) 7.407   (e) 7.42   Answer_____

B. Which number is the largest?

1. (a) .94   (b) .99   (c) .92   (d) .97   (e) .96   Answer_____

2. (a) .41   (b) .44   (c) .4   (d) .49   (e) .47   Answer_____

3. (a) 7.1   (b) 7.07   (c) 7.21   (d) 7.04   (e) 7.39   Answer_____

4. (a) 4.7   (b) 4.07   (c) 4.75   (d) 4.72   (e) 4.79   Answer_____

5. (a) 6.09   (b) .069   (c) 6.9   (d) 6.009   (e) 6.097   Answer_____

6. (a) .98   (b) .0991   (c) .0999   (d) .099   (e) .0909   Answer_____

Exercise 13:
Book pages 26–27

Add each of the following:

1. 8.2 + 3.9 + 7.6   Answer _____

2. 5.3 + 2.5 + 7.3   Answer _____

3. 6.85 + 12.9 + 17   Answer _____

4. 6.82 + 3.89 + 2.74   Answer _____

5. 19 + 12.1 + .08   Answer _____

6. 7.09 + 70.9 + 709   Answer _____

7. 13.49 + 134.9 + 134   Answer _____

8. 7.6 + 2.1 + 3.89   Answer _____

9. 6.74 + 12.9 + 7.2   Answer _____

10. 12.9 + 1.29 + 129   Answer _____

11. 1.2 + 2.3 + 3.4   Answer _____

12. 7.619 + 12 + 1.89   Answer _____

13. 7.2 + 12.8 + 139.7   Answer _____

14. 61.2 + 12.7 + 139.84   Answer _____

15. 17 + 1.7 + .17   Answer _____

Exercise 14:
Book pages 27–28

Subtract each of the following:

1. 12.9 − 3.2   Answer _____

2. 14.8 − 6.9   Answer _____

3. 13.1 − 7.07   Answer _____

4. 11.04 − 7.09   Answer _____

5. 6 − .078   Answer _____

6. 12 − .12   Answer _____

7. 55.7 − 10.84   Answer _____

8. 202.8 − 2.028   Answer _____

9. 7.2 − 6.9   Answer _____

10. 13.75 − 12.79   Answer _____

11. 6.2 − 3.89   Answer _____

12. 6 − .078   Answer _____

13. 12.2 − 3.98   Answer _____

14. 7.2 − 5.87   Answer _____

15. 8.3 − 6.91   Answer _____

Exercise 15:
Book pages 28–29

A. Multiply each of the following:

1. .4 × .6   Answer _____

2. .5 × 3   Answer _____

3. 6.7 × 2.9   Answer _____

4. 2.5 × 3.7   Answer _____

5. 48 × .06   Answer _____

6. 37 × 2.4   Answer _____

7. .01 × .01   Answer _____

8. .001 × .4   Answer _____

9. .069 × .7   Answer _____

10. 290 × .51   Answer _____

11. 3.71 × 2.09   Answer _____

12. .075 × .029   Answer _____

13. 5.04 × 2.16   Answer _____

14. 17.1 × .025   Answer _____

15. .035 × .01   Answer _____

B. Verbal Problems:

1. Find the cost of 15 candy bars at $.55 per candy bar.   Answer _____

2. A book salesman sells 12 copies of a book at $8.95 per copy. What is the total cost?   Answer _____

3. How much do 8 quarts of juice cost at $1.39 per quart?   Answer _____

4. A roll of wallpaper costs $10.95. If a bedroom requires 38 rolls of paper, what is the cost to wallpaper the bedroom?   Answer _____

5. A salesman earns $7.95 per hour he works. If he works 29 hours, how much does the salesman earn?   Answer _____

6. Pears sell for $.99 per pound. What is the cost of 7 pounds of pears?   Answer _____

Exercise 16:
Book pages 29–30

A. Divide each of the following:

1. 36 ÷ .6   Answer _____

2. 150 ÷ .5   Answer _____

3. 48 ÷ .04   Answer _____

4. .01 ÷ 10   Answer _____

5. 10 ÷ .01   Answer _____

6. 6,000 ÷ .60   Answer _____

7. 840 ÷ .42   Answer _____

8. 5.2 ÷ .13   Answer _____

9. 2.6 ÷ .13   Answer _____

10. 242 ÷ .22   Answer _____

11. 3.6 ÷ .6   Answer _____

12. .84 ÷ 12   Answer _____

13. .4949 ÷ 7   Answer _____

14. 4.949 ÷ .7   Answer _____

15. 1.68 ÷ .8   Answer _____

B. Verbal problems:

1. If notebooks cost $1.70 a piece, how many can you purchase for $32.30?   Answer _____

2. Pens cost $1.50 each. How many pens can you purchase for $60.00?   Answer _____

3. Chocolate candy sells for $7.20 per pound. How many pounds can be purchased for $144.00?   Answer _____

4. Zelma pays $11.60 for a tank of gasoline. If gasoline sell for $1.16 a gallon, how many gallons does she buy?   Answer _____

Exercise 17:
Book pages 31–33

A. Change each fraction to a decimal.

1. $\frac{7}{40}$ Answer _____

2. $\frac{5}{4}$ Answer _____

3. $\frac{7}{100}$ Answer _____

4. $\frac{1}{20}$ Answer _____

5. $\frac{4}{5}$ Answer _____

6. $\frac{3}{10}$ Answer _____

7. $\frac{1}{50}$ Answer _____

8. $\frac{7}{8}$ Answer _____

9. $\frac{4}{9}$ Answer _____

10. $\frac{2}{7}$ Answer _____

B. Change each decimal to a fraction (reduce answer to lowest terms).

1. .45   Answer _____

2. .04   Answer _____

3. .035   Answer _____

4. 1.4   Answer _____

5. 9.85   Answer _____

6. 7.8   Answer _____

7. .019   Answer _____

8. .055   Answer _____

9. .85   Answer _____

10. 7.69   Answer _____

C. Change each percentage to a fraction (reduce answer to lowest terms).

1. 95%   Answer _____

2. 4%   Answer _____

3. 21%   Answer _____

4. 26%   Answer _____

5. 250%   Answer _____

6. 80%   Answer _____

7. 89%   Answer _____

8. 765%   Answer _____

9. 72%   Answer _____

10. 8%   Answer _____

D. Change each percent to a decimal.

1. 49%   Answer _____

2. 8%   Answer _____

3. 245%   Answer _____

4. .9%   Answer _____

5. 7%   Answer _____

6. 65%   Answer _____

7. 539%   Answer _____

8. .09%   Answer _____

9. 3.7%   Answer _____

10. .79%   Answer _____

Exercise 18:
Book pages 33–35

A. Change each fraction to a decimal rounded to nearest tenth.

1. $\frac{3}{8}$ Answer _____

2. $\frac{2}{7}$ Answer _____

3. $\frac{4}{9}$ Answer _____

4. $\frac{2}{5}$ Answer _____

5. $\frac{1}{8}$ Answer _____

6. $\frac{4}{7}$ Answer _____

B. Change each fraction to a decimal rounded to nearest hundredth.

1. $\frac{5}{7}$ Answer _____

2. $\frac{7}{9}$ Answer _____

3. $\frac{11}{17}$ Answer _____

4. $\frac{15}{19}$ Answer _____

5. $\frac{6}{7}$ Answer _____

6. $\frac{2}{9}$ Answer _____

C. Verbal problems:

1. Multiply 3.7×4.6 and round off answer to nearest tenth.   Answer _____

2. Multiply 6.7×2.5 and round off answer to nearest tenth.   Answer _____

3. Multiply 6.75×2.9 and round off answer to nearest hundredth.   Answer _____

4. Multiply 12.7×83.1 and round off answer to nearest hundredth.   Answer _____

161

Exercise 19:
Book pages 35–37

Cost and profit problems:

1. What is the total cost of 45 pens at $.75 each?　　　　Answer _____

2. Find the total cost of 4 pounds of peaches at $.79 per pound and 6 pounds of plums at $.89 per pound.　　　　Answer _____

3. A mathematics teacher charges $25.00 for the first lesson and $19.00 for each additional lesson. What is the cost of 21 lessons?　　　　Answer _____

4. The Santiago Moving Company charges $45.00 for the first hour of work and $39.00 for each additional hour. What is the cost of a moving job that takes nine hours?　　　　Answer _____

5. A theatre group sells 425 tickets to a play written by Shakespeare. Each ticket sells for $6.00. The group spends $575.00 on rent and $179.00 in additional expenses. What is the profit?　　　　Answer _____

6. A department store buys 175 coats for $6,000.00. All the coats are sold at $79.00 each. What is the profit?　　　　Answer _____

7. Abraham Lincoln High Schools sells 745 tickets to a basketball game at $8.00 per ticket. It costs $2,100 a month to rent the stadium. Other expenses add up to $789.00. What is the profit?　　　　Answer _____

8. Hector's Superette bus 27 dozen rolls at $.89 per dozen. Only 21 dozen are sold at $1.89 per dozen. What is the profit?　　　　Answer _____

9. Elliot spends $14.37 in the Superette. How much change does he receive from a $100.00 bill?　　　　Answer _____

10. Philip bought two ice cream pops at $.85 each. How much change will he receive from a $50.00 bill?　　　　Answer _____

Exercise 20:
Book pages 37–39

Percent problems:

1. What is 15% of 75?   Answer_____

2. What is 17% of 90?   Answer_____

3. If 40% of a number is 80, what is the number?   Answer_____

4. If 25% of a number is 480, find the number.   Answer_____

5. What is 15% of 60?   Answer_____

6. What is 60% of 70?   Answer_____

7. If 40% of a number is 120, what is the number?   Answer_____

8. If 75% of a number is 375, find the number.   Answer_____

9. What is 70% of 30?   Answer_____

10. What is 16% of 90?   Answer_____

Exercise 21:
Book pages 39–40

Sales tax and percent increase:

1. A coat sells for $70.00. There is a 7% sales tax.
    What is the total price?   Answer_____

2. A dress sells for $29.00. There is a 6% sales tax.
    What is the sale tax?   Answer_____

3. Eddie buys Zelma a Movado watch for $2,500.00;
    there is an 8% sales tax. What is the total price?   Answer_____

4. Seth earns $79,000 a year. He receives a 9% increase in
    salary. What is his new salary?   Answer_____

5. Philip's variety store gross sales were $7,000,000 last
    year. The sales this year increased by 21%. What is the
    increase in sales?   Answer_____

6. Box seat tickets to Yankee games was increased by 18%.
    Last year the tickets sold from $30.00 each. What do they
    sell for this season?   Answer_____

163

Exercise 22:
Book pages 40–41

Discount and percent decrease:

1. John's weekly salary is $525.00. It is reduced by 7%. What is his new weekly salary?

   Answer_____

2. There is a 30% decrease on all luggage. Zelma bought an attaché case that originally sold for $189.00. What is the sale price?

   Answer_____

3. New York City has 9,000,000 people. It loses 8% of its population. What is the new population?

   Answer_____

4. Sally earns $35,000 a year. She has 14% of her pay deducted for taxes. What is here yearly take-home pay?

   Answer_____

5. A dress that sold for $47.95 is reduced by 7%. What is the sale price.

   Answer_____

6. Seth buys a sweater that originally sells for $95.00. It is reduced by 45%. What is the sale price?

   Answer_____

Exercise 23:
Book pages 42–44

Area, perimeter and cost:

1. Find the area of a room 9 years by 6 years.

   Answer_____

2. Find the area of a sheet that is 75 inches long and 40 inches wide.

   Answer_____

3. How much does it cost to carpet a room 12 yards by 11 yards at $9.00 per square yard?

   Answer_____

4. Find the perimeter of a rectangle whose length is 10 feet and whose width is 7 feet.

   Answer_____

5. How much fencing is needed to fence in a swimming pool 45 feet long and 30 feet wide.

   Answer_____

6. How much does it cost to fence in a rectangular garden 28 feet by 18 feet at $14.00 per foot of fencing?

   Answer_____

7. Carpeting costs $24.00 per square yard. How much does it cost to carpet a room 12 yards by 11 yards?

   Answer_____

8. Find the perimeter of a triangle whose sides are 6, 8 and 10 inches.

   Answer_____

A. Addition of sign numbers:

1. $(-3) + (-4)$ Answer_____   2. $(-6) + (-8)$ Answer_____

3. $(-4) + (-7)$ Answer_____   4. $-7(-4)$ Answer_____

5. $-7 + (6)$ Answer_____   6. $-3 -4$ Answer_____

7. $-8 + 4$ Answer_____   8. $-6 -9$ Answer_____

9. $(-7) + (-4)$ Answer_____

10. $-3 + 7 - 8 + 6 - 7 + 4$ Answer_____

11. $-2 + 4 - 3 + 6 - 1 + 7$ Answer_____

B. Subtraction of sign numbers:

1. $(-7) - (-4)$ Answer_____   6. $7 - (-8)$ Answer_____

2. $(-3) - (1)$ Answer_____   7. $4 - (5)$ Answer_____

3. $(4) - (7)$ Answer_____   8. $7 - (-8)$ Answer_____

4. $-7 - (-5)$ Answer_____   9. $3 - (-7)$ Answer_____

5. $-6 - (4)$ Answer_____

C. Multiplication of sign numbers:

1. $(4)(-5)$ Answer_____   5. $-7(4)$ Answer_____

2. $6(-4)$ Answer_____   6. $-3(4)$ Answer_____

3. $-7(-3)$ Answer_____   7. $(-7)(-2)(-3)(-1)(-1)(-4)$ Answer_____

4. $(-5)(-7)$ Answer_____   8. $(-3)(-1)(-2)(-1)(-7)$ Answer_____

D. Division of sign numbers:

1. $\dfrac{-72}{-8}$ Answer_____   6. $\dfrac{-6}{-3}$ Answer_____

2. $\dfrac{36}{-6}$ Answer_____   7. $\dfrac{-16}{4}$ Answer_____

3. $\dfrac{-45}{5}$ Answer_____   8. $\dfrac{-21}{7}$ Answer_____

4. $\dfrac{-81}{-9}$ Answer_____   9. $\dfrac{-60}{-12}$ Answer_____

5. $\dfrac{16}{-8}$ Answer_____

Exercise 25:
Book pages 59–61

Order of Operations:

1. $-7(4-2)$ Answer_____   7. $(-2-3)^2$ Answer_____

2. $-3(4+7)^2$ Answer_____   8. $4+3(5+6)^2$ Answer_____

3. $-3+2(4+7)^2$ Answer_____   9. $-2(-4-5)^2$ Answer_____

4. $-4(9+1)-7$ Answer_____   10. $5(7)^2$ Answer_____

5. $7-8(6)^2$ Answer_____   11. $6-2(-3)^2$ Answer_____

6. $(-4)^2+3(-6)$ Answer_____   12. $-3(-4)^2-5(-6)^2$ Answer_____

Exercise 26:
Book pages 62–63

Interpret each of the following algebraic expressions:

1. Six more than a number Answer_____

2. A number minus eight Answer_____

3. A number decreased by six Answer_____

4. A number plus seven Answer_____

5. 8 times a number Answer_____

6. 5 less than a number Answer_____

7. A number increased by 14 Answer_____

8. 4 more than one third a number Answer_____

9. 5 more than 6 times a number Answer_____

10. 7 less than 4 times a number Answer_____

11. Find the value in cents of "q" quarters and "d" dimes. Answer_____

12. Find the value in cents of "n" nickels and "q" quarters. Answer_____

Exercise 27:
Book pages 63–65

Addition and subtraction of algebraic expressions:

1. $(4x+8y)+(7x+9y)$ Answer_____

2. Add $4a-8b$ and $5a-3b$ Answer_____

3. Add $7x^2-12x$ and $9x^3-11$ Answer_____

4. $(4a^2-7a)+(-3a+9a^2)$ Answer_____

5. Add $-6x^2+9$ and $-3x+11$ Answer_____

6. $(7b^2-6a)+(-3b^2+9a)$ Answer_____

7. Subtract $8y-4$ from $-8y-4$ Answer_____

8. Subtract $4x-5$ from $x-8$ Answer_____

9. Simplify $(-4x^2+7x+10)-(-4x^2+7x-10)$ Answer_____

10. Simplify $(-7x^2-7x-10)-(7x^2+7x+10)$ Answer_____

Exercise 28
Book pages 66–67

Multiplication of a Monomial by a Monomial:

1. $(-4x^2y)(-5xy^2)$ Answer_____

2. $(-5x^3)(-9x^2)$ Answer_____

3. $(8x^2y^3)(-3x^5y^7)$ Answer_____

4. $(8x^2y^3z)(-3xyz)$ Answer_____

5. $(-6ABC)(-2ABC)$ Answer_____

6. $(-4ab^2c)(5a^5bc^2)$ Answer_____

7. $(-7x^3y^4)(3x^4y^{12})$ Answer_____

8. $(8rst^2)(-3r^2st)$ Answer_____

9. $(4a^2b^3c^2)(-5a^2b^2c^2)$ Answer_____

10. $(-7x^2y^2z)(-3x^2y^2z^2)$ Answer_____

Exercise 29
Book pages 68–69

Multiplication of a Monomial by a Binomial:

1. $6(x-7)$ Answer_____

2. $y(y^2-7)$ Answer_____

3. $-4x(x-9)$ Answer_____

4. $-7y(y-9)$ Answer_____

5. $-3x(x^2-6)$ Answer_____

6. $-6z(3z^2-8z)$ Answer_____

7. $-5ab^2(-3a+8)$ Answer_____

8. $-8x^2y(4x^2y-2)$ Answer_____

9. $-7x(x^3-x^2)$ Answer_____

10. $-3x^2y^3(4x^5y^7-3x^3y^3)$ Answer_____

11. $-7ABC(-3A^3B^3-4A^4B^4)$ Answer_____

Exercise 30
Book pages 69–70

Multiplication of a Binomial by a Binomial:

1. $(x-7)(x-6)$ Answer_____   7. $(y-8)^2$ Answer_____

2. $(x-7)(x-9)$ Answer_____   8. $(4x-7)(4x+7)$ Answer_____

3. $(4x-5)(6x-7)$ Answer_____   9. $(2x-8)(3x+9)$ Answer_____

4. $(y-6)(y+6)$ Answer_____   10. $(2x-7)(3x+9)$ Answer_____

5. $(y+7)(y-7)$ Answer_____   11. $(2x-4)^2$ Answer_____

6. $(x-7)^2$ Answer_____   12. $(3x-7)(x+9)$ Answer_____

Exercise 31
Book pages 70–71

Multiplication of Polynomials:

1. $(y+3)(y^2+4x+5)$ Answer_____

2. $(x-5)(x^2+3x+7)$ Answer_____

3. $(3x-1)(x^2+4x-6)$ Answer_____

4. $(x+7)(x^2-9x+5)$ Answer_____

5. $(y-3)(y^2+5y+7)$ Answer_____

6. $(x-4)(x^2-x+4)$ Answer_____

7. $(5x^2+x+1)+(x^2-4x+6)$ Answer_____

8. $(3x^2+x+2)+(x^2-5x+4)$ Answer_____

Exercise 32
Book pages 71–73

Evaluating algebraic expressions:

1. Find the value of $3x-5$ when $x=-3$.   Answer_____

2. Find the value of A^2-8A when $A=-5$.   Answer_____

3. Evaluate $-b^2$ when $b=-6$.   Answer_____

4. Find the value of $4A+5B$ when $A=-6$ and $B=-7$.   Answer_____

5. Find the value of x^2+7xy when $x=4$ and $y=5$.   Answer_____

6. Find the value of $4x^3+7x^2$ when $x=-1$.   Answer_____

7. Evaluate $-3x^2$ when $x=-4$.   Answer_____

8. Evaluate $-3x^3y^2$ when $x=-2$ and $y=-3$.   Answer_____

9. Find the value of x^2+6y when $x=5$ and $y=6$.   Answer_____

10. Evaluate $-5x^2$ when $x=-6$.   Answer_____

Exercise 33
Book pages 73–74

Order of Operations Algebra:

1. $4y + y(y + 8)$ Answer_____

2. $3x + x(x - 5)$ Answer_____

3. $4x + x^2(x - 7)$ Answer_____

4. $5y + 5y^2(-y + 10)$ Answer_____

5. $7x^2 - 8x(x - 5)$ Answer_____

6. $4x^2 + 5x^2(-x - 8)$ Answer_____

7. $2x + x(x - 9)$ Answer_____

8. $3x^2 + 6x(x - 7)$ Answer_____

9. $4a^2b - 3a(ab - 7)$ Answer_____

10. $x^2y - 8x(x^2 + 3xy)$ Answer_____

11. $6y + y(y - 7)$ Answer_____

12. $4x^2 - 3x(x + 5)$ Answer_____

Exercise 34:
Book pages 75–76

Raising a number to a power:

1. $(x^5)^3$ Answer_____

2. $(y^{17})^6$ Answer_____

3. $(xy^3)^4$ Answer_____

4. $(a^2b)^4$ Answer_____

5. $(c^5d^5)^6$ Answer_____

6. $(6x^2)^3$ Answer_____

7. $(-3x^2y^4)^5$ Answer_____

8. $(3x^3y^2)^2$ Answer_____

9. $(-2x^2y^7)^6$ Answer_____

10. $(a^3b^4)^5$ Answer_____

11. $(2xy^4)^3$ Answer_____

12. $(-3xy^4)^4$ Answer_____

Exercise 35
Book pages 76–78

Dividing a monomial or a binomial by a monomial:

1. $\dfrac{8x^3}{2x}$ Answer_____

2. $\dfrac{-7x^2}{-x^2}$ Answer_____

3. $\dfrac{12b - 24}{-6b}$ Answer_____

4. $\dfrac{20y - 40}{-5}$ Answer_____

5. $\dfrac{9y^2 - 18}{-3}$ Answer_____

6. $\dfrac{6x^4 - 3x^2}{3x^2}$ Answer_____

7. $\dfrac{16x^7 - 4x^5}{-2x^2}$ Answer_____

8. $\dfrac{12x^3y^7}{-6x^2}$ Answer_____

9. $\dfrac{18x^7y^8}{-2xy}$ Answer_____

10. $\dfrac{16y^3 - 6y^2}{-2y}$ Answer_____

11. $\dfrac{18x^7 - 3x^2}{3x^2}$ Answer_____

12. $\dfrac{14x^4y^7 - 2x^5y}{-2x^3y}$ Answer_____

Exercise 36
Book pages 78–83

Factor each of the following if possible:

1. $y^2 + 7y$ Answer_____

2. $x^2 - 36$ Answer_____

3. $4x^2 - 8x$ Answer_____

4. $x^2 + 6x + 5$ Answer_____

5. $y^2 + y - 6$ Answer_____

6. $3A^2 - 48$ Answer_____

7. $x^2 - 144$ Answer_____

8. $A^2 + 6A + 4$ Answer_____

9. $6x - 24x^3$ Answer_____

10. $x^2 - 5x + 6$ Answer_____

11. $6x^3 - 54x$ Answer_____

12. $x^2 - 9x + 20$ Answer_____

13. $x^2 - 7x + 12$ Answer_____

14. $5B^3 + 45B^2 + 100B$ Answer_____

15. $20x^2y - 30x$ Answer_____

16. $49A^2 - 64B^2$ Answer_____

17. $x^2 - 5x + 4$ Answer_____

18. $x^2\ 9x + 20$ Answer_____

19. $2x^2 + 18x + 28$ Answer_____

20. $36x^2 - 81y^2$ Answer_____

21. $18x^6y^2 - 9x^5y^5$ Answer_____

22. $4x^2 - 16y^2$ Answer_____

23. $x^2 + 3x + 2$ Answer_____

24. $x^2 - 15x + 54$ Answer_____

25. $x^6 - 81$ Answer_____

26. $x^8 - 36$ Answer_____

27. $4x^4 - 81$ Answer_____

28. $7x^3y^2 - 14x^5y^2$ Answer_____

29. $x^2 + 4x - 5$ Answer_____

30. $16 - x^2$ Answer_____

Exercise 37
Book pages 83–85

Solve each of the following equations:

1. $x - 6 = 4$ Answer_____

2. $x + 8 = 6$ Answer_____

3. $-3x = 12$ Answer_____

4. $-4y = -16$ Answer_____

5. $2y - 9 = 15$ Answer_____

6. $-3x - 6 = -12$ Answer_____

7. $5 - y = 0$ Answer_____

8. $6 - 7y = 20$ Answer_____

9. $3x - 6 = 2x + 15$ Answer_____

10. $4x - 7 = x + 8$ Answer_____

11. $3x + 6 + 2x = 31$ Answer_____

12. $4x - 7 - x = 2$ Answer_____

13. $2y - 9 = 5$ Answer_____

14. $-3(x - 6) = 18$ Answer_____

15. $-2(x + 4) = -8$ Answer_____

16. $7x - 9 = x + 15$ Answer_____

17. $7x - 2 = 2x + 23$ Answer_____

18. $3(x - 7) = -3$ Answer_____

19. $3x + 4 = 2x + 6$ Answer_____

20. $2y - 6 = -12$ Answer_____

21. $3x - 7 = 2x + 5$ Answer_____

Exercise 38
Book pages 85–86

Solving equations with fractions:

1. $\frac{y}{9} + \frac{y}{3} = 4$ Answer_____

2. $\frac{y}{2} + 3 = \frac{y}{4}$ Answer_____

3. $\frac{x}{3} + 2 = 6$ Answer_____

4. $\frac{x}{5} = \frac{x}{10} - 4$ Answer_____

5. $\frac{y}{3} + \frac{y}{2} = 5$ Answer_____

6. $\frac{y}{24} + 2 = \frac{y}{20}$ Answer_____

7. $\frac{x}{3} + 1 = \frac{x}{2} + 6$ Answer_____

8. $\frac{y + 1}{4} = \frac{3y + 6}{2}$ Answer_____

9. $\frac{x + 7}{4} = \frac{3x - 6}{2}$ Answer_____

10. $\frac{x}{3} + \frac{x}{4} = 12$ Answer_____

11. $\frac{y - 7}{3} = \frac{2x + 7}{4}$ Answer_____

12. $\frac{x}{4} - 7 = 14$ Answer_____

Exercise 39
Book pages 87–88

Solving literal equations:

1. $x-y=9$ Solve for x: Answer_____

2. $a+8b=0$ Solve for a: Answer_____

3. $4x-5=y$ Solve for x: Answer_____

4. $5x-6y=7$ Solve for y: Answer_____

5. $\frac{x}{3} + 4 = y$ Solve for x: Answer_____

6. $\frac{x}{4} - 8 = y$ Solve for x: Answer_____

7. $x-2y=0$ Solve for y: Answer_____

8. $A+B+C=0$ Solve for B: Answer_____

9. $4x-5y+8z=3$ Solve for x: Answer_____

10. $4(x+y)=6$ Solve for y: Answer_____

11. $\frac{x - y}{3} = c$ Solve for a: Answer_____

12. $\frac{A - B}{2} = d$ Solve for a: Answer_____

13. $\frac{x}{4} - 8 = y$ Solve for x: Answer_____

14. $2a-3b=4c$ Solve for a: Answer_____

Exercise 40
Book pages 89–90

Solve the following problems:

1. The sum of four times a number and seven is sixty-seven. Find the number.

   Answer_____

2. One number is six less than another. Their sum is twelve. Find both numbers.

   Answer_____

3. The sum of twice a number and sixteen is twenty-four. Find the number.

   Answer_____

4. Six times a number and seven is forty-three. Find the number.

   Answer_____

5. One number is eight less than another. The sum is twenty-four. Find the larger number.

   Answer_____

6. A number divided by ten is thirty. Find the number.

   Answer_____

7. Two more than five times a number is thirty-seven. Find the number.

   Answer_____

8. One number is ten less than another, their sum is fifty. Find the smaller number.

   Answer_____

9. At a baseball game, each ticket costs $10. If expenses total $3,000, how many tickets must be sold in order to have a profit of $2,000?

   Answer_____

10. It costs a manufacturer $20 to produce each video game. In addition, there is a general overhead of $30,000. He produces 6,000 games. If he received $50 per game from a wholesaler, how many games must he sell to break even?

    Answer_____

Exercise 41
Book pages 91–92

Solve each of the following proportions:

1. $\frac{x}{4} = \frac{3}{2}$ Answer_____

2. $\frac{1}{2} = \frac{x}{4}$ Answer_____

3. $\frac{2}{9} = \frac{x}{18}$ Answer_____

4. $\frac{2}{7} = \frac{x}{21}$ Answer_____

5. $\frac{3}{x} = \frac{1}{9}$ Answer_____

6. $\frac{2x}{7} = \frac{4}{9}$ Answer_____

7. $\frac{5 + x}{3} = \frac{1}{6}$ Answer_____

8. $\frac{2x + 7}{3} = \frac{3}{8}$ Answer_____

9. $\frac{x - 7}{2} = \frac{1}{4}$ Answer_____

10. $\frac{x}{7} = \frac{2}{7}$ Answer_____

11. $\frac{3x - 1}{4} = \frac{9}{2}$ Answer_____

12. $\frac{3x}{4} = 15$ Answer_____

Exercise 42
Book page 93

Solve each of the following verbal problems by the method of ratios and proportions:

1. If 400 grams of candy contains 68 grams of fat, how many grams of fat are there in 700 grams of candy?   Answer_____

2. Zelma drives her car 720 miles in 8 hours. At this rate, how far will she travel in 11 hours?   Answer_____

3. A map is drawn so that 2 inches represents 700 miles. If the distance between 2 cities is 3,850 miles, how far apart are they on a map?   Answer_____

4. A 7-ounce serving of orange juice contains 105 grams of water. How many grams of water are in a 12-ounce serving of orange juice?   Answer_____

5. A baseball player makes 2 hits every 9 times at bat. If he hits the ball with the same accuracy, how many hits would he get if he were at bat 450 times?   Answer_____

Exercise 43
Book pages 94–98

Solve each of the following simultaneous equations:

1. $x - y = 7$
   $x + y = 1$ Answer_____

2. $x + 3y = 4$
   $2x - y = 8$ Answer_____

3. $2x + 4y = 6$
   $x = y + 3$ Answer_____

4. $-3x + 2y = 6$
   $3x + y = 3$ Answer_____

5. $A + 2B = 6$
   $A - 3B = 1$ Answer_____

6. $3A - 4B = 0$
   $A = 7 - B$ Answer_____

7. $3A + B = 10$
   $2A + 3B = 9$ Answer_____

8. $A + B = 6$
   $2A - B = 3$ Answer_____

9. $2x + 3y = 9$
   $3x + y = 10$ Answer_____

10. $2A - B = 6$
    $3A + B =$ Answer_____

11. $3x + y = 6$
    $y = x + 2$ Answer_____

12. $4A + 3B = 2$
    $A + B = 0$ Answer_____

13. $2A - B = 7$
    $A + B = 2$ Answer_____

14. $3x - y = 4$
    $y = x$ Answer_____

15. $2x + 5y = 9$
    $4x - 3y = 5$ Answer_____

16. $x - 2b = 10$
    $x + 2b = 6$ Answer_____

17. $4A - B = 12$
    $A + 2B = 30$ Answer_____

18. $4x + 8y = 12$
    $x = y$ Answer_____

19. $-4x + 2y = 12$
    $4x + 4y = 12$ Answer_____

20. $2A + 3B = 7$
    $A = B - 4$ Answer_____

21. $4x + 12y = 8$
    $2x + 4y = 6$ Answer_____

Exercise 44
Book pages 98–99

Solve each of the following quadratic equations:

1. $x^2+9x=-20$ Answer_____

2. $y^2-4y=-3$ Answer_____

3. $x^2+9x=-8$ Answer_____

4. $x^2=36$ Answer_____

5. $x^2-8x=-15$ Answer_____

6. $x^2-x=20$ Answer_____

7. $x^2+3x=4$ Answer_____

8. $x^2+8x=-7$ Answer_____

9. $x^2+6x=-8$ Answer_____

10. $A^2+5A=-6$ Answer_____

11. $x^2+9x=-8$ Answer_____

12. $A^2-7A=18$ Answer_____

13. $y^2-3y=-2$ Answer_____

14. $y^2-6y=16$ Answer_____

15. $x^2-9x=-14$ Answer_____

Exercise 45
Book pages 100–102

Solve the following verbal problems:

1. Which one of the following points lies on the graph: $y=2x+1$
   (a) $(-3, 7)$ (b) $(1, 3)$ (c) $(2, 9)$ (d) $(3, 4)$ Answer_____

2. Which one of the following points lies on the graph: $2x+y=5$
   (a) $(-3, -5)$ (b) $(4, 1)$ (c) $(2, 1)$ (d) $(-7, -5)$ Answer_____

3. Which one of the following points lies on the graph: $y=x-1$
   (a) $(2, 9)$ (b) $(6, -5)$ (c) $(3, -4)$ (d) $(4, 3)$ Answer_____

4. Which one of the following points lies on the graph: $7x-6y=1$
   (a) $(3, 7)$ (b) $(1, 1)$ (c) $(-2, -7)$ (d) $(-3, -4)$ Answer_____

5. Which one of the following points lies on the graph: $x-y=2$
   (a) $(4, 2)$ (b) $(3, 7)$ (c) $(-6, -9)$ (d) $(2, 9)$ Answer_____

6. Which one of the following points lies on the graph: $y=5-x$
   (a) $(2, 7)$ (b) $(6, 9)$ (c) $(-3, -7)$ (d) $(3, 2)$ Answer_____

7. Which one of the following points lies on the graph: $6x+2y=0$
   (a) $(-3, -7)$ (b) $(2, 7)$ (c) $(1, -3)$ (d) $(-2, 9)$ Answer_____

Exercise 46
Book pages 102–103

Solve the following verbal problems:

1. Find the x intercept for the equation $3x - y = 9$: Answer_____

2. Find the y intercept for the equation $3x - 7y = -14$: Answer_____

3. Find the x intercept for the equation $3x - 7y = 12$: Answer_____

4. Find the y intercept for the equation $-6x - 9y = 18$: Answer_____

5. Where does the graph $3x + 7y = -15$ intercept the x-axis? Answer_____

6. Where does the graph $-4x - 6y = -12$ intercept the y-axis? Answer_____

7. Where does the graph $3x + 6y = 24$ intercept the x-axis? Answer_____

8. Where does the graph $7x - 9y = 18$ intercept the y-axis? Answer_____

Exercise 47
Book pages 104–105

For each of the following, write an equation for the line:

1.
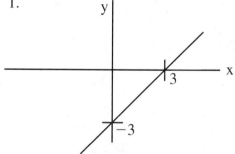

2.
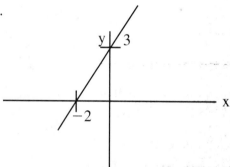

3.
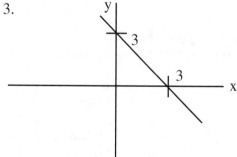

4.
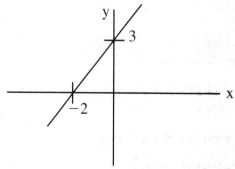

179

Exercise 48
Book pages 106–107

Find the slope for each of the following sets of points:

1. (3, 2) (5, 6)   Answer_____

9. (−2, 1) (−4, 8)   Answer_____

2. (3,) (5, 9)   Answer_____

10. (−3, −7) (−4, −1)   Answer_____

3. (−6, −2) (−4, −9)   Answer_____

11. (−3, 7) (4, −9)   Answer_____

4. (−9, 2) (−3, 4)   Answer_____

12. (4, −9) (−3, 7)   Answer_____

5. (7, −6) (−3, 4)   Answer_____

13. (7, 1) (1, 7)   Answer_____

6. (−3, −7) (28, −1)   Answer_____

14. (−3, −4) (−4, −3)   Answer_____

7. (2, 5) (7, 9)   Answer_____

15. (6, 1) (7, 4)   Answer_____

8. (3, −6) (−2, 9)   Answer_____

Exercise 49
Book pages 107–108

For each of the following problems, write an equation:

1. m=5, b=2   Answer_____

6. m=2, b=3   Answer_____

2. m=3, b=−7   Answer_____

7. $m = \frac{-2}{3}, b = 9$ Answer_____

3. m=−2, b=5   Answer_____

8. $m = \frac{-3}{5}, b = \frac{1}{4}$ Answer_____

4. m=−5, b=−7   Answer_____

9. $m = \frac{2}{9}, b = \frac{4}{5}$ Answer_____

5. m=−2, b=−1   Answer_____

Exercise 50
Book pages 108–109

Find the slope and the y-intercept for each of the following equations:

1. $x + 4y = 8$ Answer_____

2. $y + 3x = 12$ Answer_____

3. $y = -3x + 6$ Answer_____

4. $x = 6 - y$ Answer_____

5. $2x + 3y = 12$ Answer_____

6. $2y + 4x = 8$ Answer_____

7. $x - y = 4$ Answer_____

8. $y = 9 - 3x$ Answer_____

9. $3x - 7y = 21$ Answer_____

10. $y = x$ Answer_____

11. $6 = 3x + y$ Answer_____

12. $7 = -x - y$ Answer_____

Exercise 51
Book pages 112–113

Answer each of the following questions:

1. Find the complement of an angle that is 61°.   Answer_____

2. Find the supplement of an angle that is 121°.   Answer_____

3. An angle is eight times its supplement. Find the smaller angle.   Answer_____

4. An angle is five times its complement. Find the larger angle.   Answer_____

5. An angle is 80° more than its supplement. Find both angles.   Answer_____

6. An angle is 24° less than its complement. Find the smaller angle.   Answer_____

7. An angle exceeds its supplement by 120°. Find the larger angle.   Answer_____

8. An angle is twice its complement. Find the larger angle.   Answer_____

Exercise 52:
Book pages 114–115

Answer each of the following verbal problems:

1. The vertex angle of an isosceles triangle is 50°. Find the value of each base angle.

    Answer_____

2. The three angles of a triangle are in the ratio 2: 3: 4. Find the value of the larger angle.

    Answer_____

3. The vertex angles of an isosceles triangle is 40°. Find the value of each base angle.

    Answer_____

4. The vertex angle of an isosceles triangle is three times the base angle. Find the value of the vertex angle.

    Answer_____

Exercise 53:
Book pages 115–116

Answer each of the following verbal questions:

1. Find the circumference of a circle whose radius is 22 cm.

    Answer_____

2. Find the radius of a circle whose circumference is 28 cm.

    Answer_____

3. Find the area of a circle whose diameter is 28 cm.

    Answer_____

4. Find the circumference of a circle whose radius is 14 cm.

    Answer_____

5. Find the radius of a circle whose circumference is 44 cm.

    Answer_____

6. Find the area of a circle whose radius is 16 cm.

    Answer_____

Exercise 54:
Book pages 117–120

Answer each of the following verbal problems:

1. Find the perimeter of a rectangle whose dimensions are 14 cm. and 12 cm.

   Answer_____

2. If the perimeter of a square is 60 cm., what is the value of the side of the square?

   Answer_____

3. The area of a triangle is 60-sq. cm. If the base is 12 cm., find the value of its altitude.

   Answer_____

4. Find the side of a square whose perimeter is 48 ft.

   Answer_____

5. The area of a trapezoid is 144-sq. cm. And the sum of the bases is 72 cm. Find the altitude.

   Answer_____

6. Find the area of a rectangle whose perimeter is 60 ft. and whose length is 20 ft.

   Answer_____

7. Find the perimeter of a rectangle whose area is 300-sq. ft. and whose length is 125 feet.

   Answer_____

8. Find the area of a rectangle whose length is 30 cm. and whose width is 12 cm.

   Answer_____

Exercise 55:
Book pages 120–121

Find the length of the missing side in each of the following problems:

1.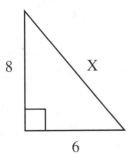

2.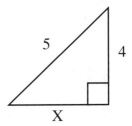

3.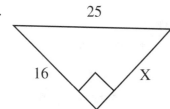

4.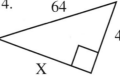

5.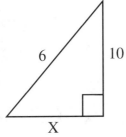

6.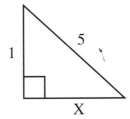